Tim James is a secondary-school science teacher, YouTuber, blogger and Instagrammer. Raised by missionaries in Nigeria, he fell in love with science at the age of fifteen and refuses to get over his infatuation. After graduating with a Master's degree in chemistry, specialising in computational quantum mechanics, he decided to get straight into the classroom. His first book, *Elemental: How the Periodic Table Now Explains (Nearly) Everything* was published in 2018.

Also by Tim James

Elemental

Fundamental

*How quantum and particle physics explain
absolutely everything (except gravity)*

TIM JAMES

ROBINSON

ROBINSON

First published in Great Britain
in 2019 by Robinson

9 10 8

Copyright © Tim James, 2019

Epigraph excerpt on p. ix
from *Nemesis* by Isaac Asimov
(New York: Bantam Books, 1990)

A CIP catalogue record for this book
is available from the British Library.

ISBN: 978-1-47214-348-8

Typeset in Scala by Hewer Text UK Ltd
Printed and bound in Great Britain by
Clays Ltd, Elcograf S.p.A.

Papers used by Robinson are from well-
managed forests and other responsible sources.

Robinson
An imprint of
Little, Brown Book Group
Carmelite House
50 Victoria Embankment
London EC4Y 0DZ

An Hachette UK Company
www.hachette.co.uk

www.littlebrown.co.uk

Dedicated to
the students of Northgate High School

Contents

'No matter how sure scientists think they are, nature has a way of surprising them.'

Nemesis by Isaac Asimov

The End

Nature is out of her mind. When you get right down to the fundamental laws of physics, right down to the basement, you find yourself in a realm of craziness and chaos where knowledge and imagination become the same thing.

This should not come as a surprise of course – you have to question the sanity of a universe which permits the existence of starfish – but even if you are prepared for nature to be eccentric, nothing braces you for quantum physics.

It began at the end of the nineteenth century when everyone was feeling smug about themselves. We had mapped the stars, isolated DNA and were on the verge of splitting the atom. Our knowledge was nearly complete and it looked as though we were about to witness the grand finale of human achievement: the end of science itself.

There were obviously a few awkward scientific puzzles nobody had quite solved but they were minor curiosities dangling to one side, like loose threads hanging off a tapestry. It was only when we gave these threads a tug that the whole picture we had been weaving for centuries began to unravel and we were brought face to face with a new picture of reality. A quantum one.

The Nobel Laureate Richard Feynman once opened a series of lectures on quantum physics by saying: 'My physics students do not understand it. I do not understand it. Nobody does.'[1] These are sobering words to hear from arguably the greatest quantum physicist in history. After all, if someone as brilliant as Feynman could not fold his brain around the topic, what chance do the rest of us mortals have?

What has to be appreciated, however, is that Feynman was *not* saying quantum physics is too complicated to understand. He was saying quantum physics is too darn strange.

Suppose someone told you to picture a four-sided triangle, or to think of a number that is smaller than ten but bigger than a billion. Those instructions are not complicated but you could not easily follow them because they are nonsensical. This is what our journey into quantum physics is going to look like.

It is a world of four-sided triangles and numbers that do not follow ordinary rules; a place where parallel universes and paradoxes lurk around every corner and objects do not have to pay attention to space or time.

Unfortunately our brains are not built to handle this kind of madness and the words we have at our disposal are not weird enough to capture nature as she truly is. That is why the physicist Niels Bohr said that when it came to quantum physics 'language can be used only as poetry'.[2]

The mistake a lot of people make is to find the whole thing baffling and decide they are not clever enough to grasp it. Do not be troubled though. Frankly, if you find this subject bizarre and unsettling that puts you alongside the greatest minds in history.

CHAPTER ONE

Glowing with Pride

Some Light History

Quantum physics began with trying to understand light, something we have been scratching our collective heads over for millennia. The Greek philosopher Empedocles, some time around the fifth century BCE, was the first person to theorise what light is.

He believed the human eye contained a magical fire-stone, which shone rays outward from our faces, illuminating whatever we wanted to look at.[1] A poetic idea, but with an obvious flaw: if our eyes are generating the light we should always be able to see in the dark because our eyes themselves are torches.

Empedocles was also the guy who gave us the now debunked idea of four elemental substances (fire, water, wind and earth) as well as attempting to explain biological diversity as the result of bodiless limbs crawling around the world until they joined up with each other at random to form animals.

Really, Empedocles's job in scientific history was to come up with bonkers ideas everyone else proved wrong. Although in the case of light rays, it took us about thirteen hundred years to realise his mistake.

It was not until the Arabic scholar Alhazen came along that we finally let go of Empedocles's notion. Alhazen carried out an experiment in which he dissected a pig's eyeball and showed that light bounced around inside the cavity the same way it does in a dark room, i.e. light is coming from objects around us and our eyes just happen to intercept their paths.[2]

It might seem weird that it took over a thousand years for us to be sure our eyes were not zapping out magical lasers, but those were different times. Back then everyone assumed humans gave objects their purpose for existing, so there was no need for them to have an appearance when they were not being looked at.

7

Fortunately, Alhazen's suggestion that experiment should trump human ego gradually caught on and we decided that light, whatever it was, came from the objects themselves and entered our eyes in straight lines. Cue the Renaissance.

Arguably, the most influential Renaissance scientist/philosopher was René Descartes, who gave us our next big idea about the physics of light.

Descartes noticed that when a candle is lit, the illumination can simultaneously reach every corner of a room, the same way a ripple started in the centre of a pond can reach every edge at the same time. Light, he reasoned, was a similar phenomenon; there was an invisible material surrounding us in every direction, which he called plenum, and light was the result of ripples and waves moving through it.[3]

The only person to disagree with his plenum-wave idea was Isaac Newton, who chiefly made it his business to disagree with anyone he considered less intelligent than himself (which was basically everyone).

Newton pointed out that if light was a wave moving through a medium it should bend around an object as it went past, the way a water wave will curve slightly as it goes around a rock. This would give shadows blurry edges, but since they are sharply defined it made more sense to think of light being made of particles, which he called 'corpuscles'.[4]

The corpuscular light theory was inevitably accepted over Descartes's plenum waves, largely due to Newton's celebrity status and the fact that he was a bully to anyone who challenged him.

Newton would have been aghast, therefore, to hear the results of an experiment carried out by a man named Thomas Young, which showed the opposite conclusion, seventy years after his death. By that, I mean the experiment was carried out seventy years after *Newton*'s death. Thomas Young did very few experiments after his own.

THE TALENTED MR RIPPLE

Thomas Young possessed one of the most remarkable minds of the eighteenth century. He is probably best known for translating the Rosetta stone and thus becoming the first modern man to decipher Egyptian hieroglyphs. He was also the first person to recognise colour receptors in our eyes, wrote several books on medicine, spoke fourteen languages, played a dozen instruments and developed our modern theory of elasticity.[5]

The experiment of his that really caused waves for light theory (pun very much intended) was one he performed in 1803, known as the double-slit experiment.

Let's go back to the idea of waves moving across a pond for a moment. Imagine a regular pulse of waves moving over a calm liquid surface and passing through a barrier with a gap in it. As these waves waft to the other side of this gap they fan out slightly – a process we call diffraction.

The reason they spread out is because the edge of a wave dissipates its energy to the surrounding water. Viewed from above, we get a pattern looking like the one below, where wave peaks are drawn as solid lines and wave troughs are dashed:

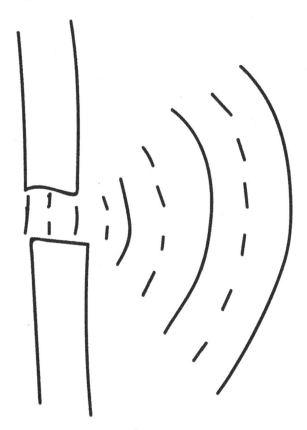

Now let's try it with two gaps in our barrier instead. The same thing will happen, only this time we see two waves diffracting through at the same time, eventually to the point where they overlap and mix together. Viewed from above it looks like this:

In some places you can see the waves are crossing over perfectly, with a peak from one wave meeting a peak from the other, leading to a mega-peak in the surface of the water. In between these mega-peaks we get the opposite effect, where the waves are out of sync and peak meets trough. In those spots the waves cancel out, leaving hardly any wave at all.

If we were to place a screen at the end of the pond now, the mixed up waves would strike it in alternating patches of mega-peak and cancelled-out nothingness. Looking at this screen head on (rather than top down) the pattern left by our waves appears like this:

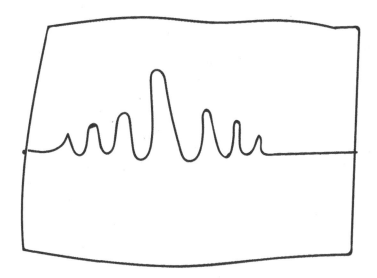

We are looking here at the effect of waves interfering as they diffract through a double slit, creating a pattern of alternating high intensity and low intensity on the other side. A phenomenon we call wave 'superposition'.

What Thomas Young then did was to replicate this wave superposition pattern, only with light beams instead of water. By shining a candle through two slits in a wall, Young ended up creating alternating zebra stripes of light and shadow on his detector screen, similar to the pattern left by mixing water waves:

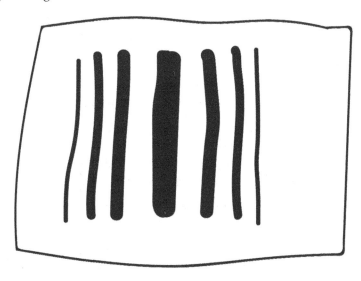

If light is made of particles as Newton insisted, they ought to shoot through the two slits and hit the wall in one big mush on the other side. The zebra pattern we actually get can only be explained if light is, in some way, wavelike.

Newton's sharp-edged-shadow objection still held some sway, but now that he was dead a few people were daring to question his teachings. If you look at the boundary of a shadow really closely you actually *do* get blurry edges: they're just small and easy to miss. This cannot be explained with a particle theory but can be explained as a wave bending around the object.

The material carrying these waves, which Descartes had called plenum, was given a fancier name – luminiferous aether – and the nature of light was finally settled.

Descartes's idea was definitely ahead of its time but it was not accepted until there was experimental proof. This is a powerful reminder that you cannot put Descartes before the horse. I'm almost sorry for that joke. Almost.

Catastrophe of the Century

By the time the 1900s rolled around, nobody was questioning what light was made of any more. Young had settled it. There were a few things that did not add up though, the most notable being what happened when light interacts with a hot object and in order to understand this mystery, we're going to have to talk about hosepipes.

Imagine a hosepipe whose spout is plugged into the bottom of a box. When we switch the hose on, the box will gradually fill with water until it cannot hold any more. But suppose we cut three holes into the lid – one small, one medium, one large.

When we switch the hosepipe on this time, the water will fill up as usual, but then begin pouring out of the holes in the top. Clearly we'll get the most water coming from the biggest hole and only a meagre trickle coming from the smallest. This would be a slightly pointless contraption to build, but there is nothing difficult going on. We pump water in at the bottom, it spills out of holes at the top.

This is a fairly good way of visualising why an object glows as it gets hotter. As any object warms, the heat energy gets absorbed and absorbed until the object has taken enough, at which point it begins leaking back out in the form of light.

The hosepipe in the analogy represents heat being applied to the object and the holes stand for different types of light that can be emitted. The smallest hole represents infrared (too low in energy to see), the middle hole represents visible (red through violet) and the largest hole represents ultraviolet (too high in energy to see).

Dark-coloured objects tend to do this heat–light conversion process most efficiently since they absorb all the energy hitting them, and thus a theoretically perfect heat-absorber is called a 'black body' in physics jargon (even if it is not literally a black object).

The whole thing is described adequately by a simple equation called the Rayleigh–Jeans law and it works as a good approximation, especially at cold to moderate temperatures. But when things get blazing hot, something really odd happens.

Logically, most of the light emitted from a hot object should emerge in the form of ultraviolet because it's the highest energy light (the biggest hole in our box analogy). What actually happens is that almost all the light comes out of the object with a medium value.

You get a little infrared and a little ultraviolet, but most of the light emitted from a hot object splurges out as yellow/orange, which does not make any sense. It would be like filling our box with water and having it all spout out from the middle hole, rather than the big one.

In fact, the real situation is even more perplexing than our three-hole analogy, because real light can have any energy it likes, rather than being limited to only three types. A more accurate picture might be to imagine a slit cut along the top of the box and finding water gushing from the middle of the slit only, somehow ignoring the edges.

The physicist Paul Ehrenfest referred to this conundrum as a 'catastrophe of the ultraviolet'[6] and it has ever since been referred to in physics books as the infamous 'ultraviolet catastrophe'.

What we have in this situation is a mismatch between theory and experiment, and in science it's always the theory that has to change. You do not get to tell an experiment what results it should produce, so if your theory does not predict the data you actually get, it's goodbye to your theory.

The catastrophe arose because we apparently had some incorrect ideas about how light energy works but nobody could have guessed that slightly rethinking those ideas would put us on a path to the quantum revolution. Although the man who came up with the answer was not trying to do anything so radical. He just wanted a cheap light bulb.

Before Plancking Was a Thing

Max Planck was the youngest of six children and graduated high school in 1875, a year ahead of his classmates. He applied to study physics at the University of Munich but the man who considered his application, Professor P. von Jolly (genuine name), tried to dissuade him because physics was almost complete and it would be a waste of Planck's intellect.[7]

Despite Jolly being so serious, Planck did not back down and insisted he be allowed to study the course he wanted. He did not care if he discovered anything new because he was not concerned with legacy. He just wanted to understand how the world worked and would not take no for an answer. Planck did not bend.

Jolly was so impressed with this dogged attitude that he decided to admit Planck after all and he soon became one of the most revered figures on the European physics circuit. His lectures were allegedly so popular people would cram shoulder to shoulder to hear him speak and there are reports of punters fainting from the heat and everyone else ignoring them so Planck could finish what he was saying.

It was this reputation that brought him to the attention of the German Bureau of Standards, which asked if he would help in its quest to provide Germany with electric streetlighting. Electricity was all the rage in other countries but it was expensive and they wanted to figure out the most efficient way of getting it done. Planck accepted the invitation gladly and set to work analysing the relationship between heat and light for a hot bulb.[8]

The filament of such a bulb is effectively a 'black body'. As it heats up from the inside the outer surface absorbs all the energy and re-emits it as light, mostly of the visible variety. As it gets hotter, however, it does not start producing the kinds of light predicted by the Rayleigh–Jeans law, so Planck decided to invent a new law, one in which he thought of light energy as a kind of gas.

In a gas, you've got a group of particles flying about at random and as they collide their heat gets shared out. By sheer chance some particles will have a low energy and some will have a high energy, but most of them will converge on an average value – what we call the temperature.

Planck realised that this distribution of energy matched what he was seeing in his light-bulb experiments. As we heat an object the emitted light hovers around a middle energy value with a few beams coming out at the high and low ends. Planck therefore proposed that energy gets shared among beams of light the same way heat gets shared among particles in a gas.

The only catch is that the gas–heat phenomenon only happens because a gas is split into particles. If Planck's idea was to work then light would have to be made of particles too.

He called these tiny granules of light 'quanta' from the Latin word *quantitas*, which means quantity, and carried on with his work unperturbed.

To be clear, Planck was not sincerely claiming light was made of particles – that would be absurd. He was just pulling a silly maths trick, largely out of desperation, to make his results sensible. Everyone knew light was a wave moving through the luminiferous aether due to Young's experiment; we had got rid of Newton's corpuscle idea a long time ago.

As far as Planck was concerned, light quanta were a halfway answer not to be taken seriously. So naturally, when he received a research paper proving they were real things he was stunned. Stiff as a plank in fact.

CHAPTER TWO

Bits and Pieces

DOCTOR WHOM?

Planck had largely forgotten about his light-particles idea by 1905 and was working as a senior editor for *Annalen der Physik*, one of the most prestigious physics journals in the world. Being in that role meant he got a lot of quack suggestions delivered to his in-tray, most of which he discarded.

The essay he received in March of that year, claiming that light really was made of particles and it wasn't just a fudge to make the numbers work, seemed like another lunatic idea at first. It came from an unknown twenty-six-year-old amateur physicist from Switzerland who boasted a qualification for teaching high school and little else. Yet the physics in the paper was not only flawless, it solved another puzzle people had been trying to answer for years.

Planck could not believe it at first and even sent his assistant to Switzerland to check if this 'A. Einstein' fellow was real and not someone writing under a pseudonym to avoid ridicule.[1] When he discovered that Einstein was in fact a genuine human (although a fairly inexperienced one – he did not even have a doctorate) Planck published his paper at once. His ridiculous light-quanta idea might not be so ridiculous after all.

Einstein's paper dealt with something called the photoelectric effect. Simply put: when you shine light on a clean piece of metal, electrons on the outside of the metallic atoms can get dislodged and come flying away from the surface.

The reason it happens is because electrons absorb light and, if the incoming beam is energetic enough, an electron can absorb it and be shaken loose. This is not that surprising in itself, but what *is* surprising is that not every colour makes it happen.

Each metal is unique but generally speaking red, orange and yellow lights do nothing to a metal surface, whereas green, blue and violet will cause electron emission. Green, blue and violet light packs more energy than red, orange and yellow so that makes sense, but what is strange is that if you increase the brightness of a red light (until it equals a blue one) nothing happens.

We measure quantum energy in units called electron-Volts (eV for short) and a red light of 10 eV contains the same amount of energy as a blue light of 10 eV. So how come red and blue light of equal energy do not cause the same effect? Is 10 eV of red not the same as 10 eV of blue? Einstein showed that if you took Planck's quantum theory seriously, it was not. Ten does not always equal ten.

An Apple in the Hand Is Worth Two Nobel Prizes

Imagine someone holding an apple in an outstretched arm. If you spray a water pistol at their hand (do not ask why you are doing this, this is how physics analogies work) the apple will stay put until you crank things up to a powerful stream. At this point, the water energy can overcome the hand's grip and the apple goes flying into the air.

In the same way, an electron is bound to its atom with a certain amount of energy and as we increase the brightness of the incoming light, we should eventually knock it loose, no matter what colour we're dealing with. We do not get that result in the lab, however, so once again we need to crumple up our theory and try something new.

If we consider our red beam and imagine chopping it into little pieces as Planck suggested, each one will contain a certain amount of energy. A beam of blue can be chopped into the same number of pieces, of course, but each one will pack more of a punch.

Instead of thinking of light energy as a smooth water jet, we need to think in terms of particles. Red light quanta would be akin to something like ping-pong balls. If you fire a ping-pong ball at someone's hand, the apple will not budge even if you boost the intensity. You can chuck a whole bucket of ping-pong balls at your volunteer but because each interaction between apple and ball is insignificant, nothing ever gets displaced no matter how much light there is.

By contrast, a quantum of blue light is more like a cannonball. If you fire a single blue particle at your apple holder it will dislodge the apple and probably their hand as well. Pick your volunteer carefully.

A hundred ping-pong balls might have the same *overall* energy as a single cannonball but the cannonball is easily going to have more of an impact. Therefore, the total energy of a beam of light is irrelevant if it is split into particles; all that matters is the colour. Which is what we observe.

According to Einstein, Planck's quantum theory of light had real physical meaning. It was not just a way of getting us near the answer, it was the literal answer itself. Light was made of particles after all. Soon after Einstein's proof, the chemist Gilbert Lewis decided these particles ought to have a catchier name than 'light-quanta' and started using the word *photon* (Greek for light), which has now stuck.[2]

Both Planck and Einstein received Nobel Prizes for their new approach to light physics in 1918 and 1921 respectively. Gilbert Lewis did not get a Nobel Prize, sadly, but he did have an awesome moustache and is credited with having invented the word 'jiffy', so in a way everyone was a winner.

UM . . . EINSTEIN? WE HAVE A PROBLEM

When Einstein proved light was made of particles it was not just confirming Planck's quantum theory, it was flying in the face of Young's light waves as well.

On the one hand, the photoelectric effect and ultraviolet catastrophe could only be explained if light was made of particles. On the other hand, the double-slit experiment shows light has to be a wave in some sort of background medium.

When two proposed hypotheses clash, scientists resolve the

disagreement by carrying out experiments to distinguish them. But what in the name of Newton's apple-munching ghost do we do when the experiments themselves disagree? This was an unprecedented situation for science so we had to try and find a loophole.

Perhaps we could explain the results of Young's double-slit experiment in terms of photons. Is our light source spraying them out like a machine gun and they collide mid-air to generate the zebra pattern?

The best way to confirm this would be to eliminate the possibility of photons interacting with each other as they fly through the double slits. Rather than spraying them all in one go we should try firing them individually, effectively replacing our machine gun with a sniper rifle.

Lots of versions of this experiment have been devised over the years but the best, hands down, was the one carried out by Akira Tonomura in 1994 while working for Hitachi.[3] The same company that make tanks, refrigerators and massage wands also holds claim to the most precise double-slit experiment ever executed.

The details of Tonomura's setup are quite different to the one Thomas Young carried out but they achieve the same goal so for simplicity and convenience I am going to use the same terminology, even though it was not quite as simple as I am going to make it sound.

In his experiment, Tonomura's beam emitter could be modified from high to low brightness, firing photons towards two slits. A detector screen was set up on the other side made from a material that illuminated when struck, creating a pinprick of light wherever each particle landed.

When a whole bunch of light was fired at the slits, as Young did in his original, Tonomura got the expected zebra pattern, but when he dialled down the intensity to one photon at a time he got something seriously weird.

For the first few minutes nothing interesting happened. Each photon shot toward the slits and hit the detector screen seemingly at random. But as he watched, the pattern of dots started to build up in ribbons like this . . . look familiar?

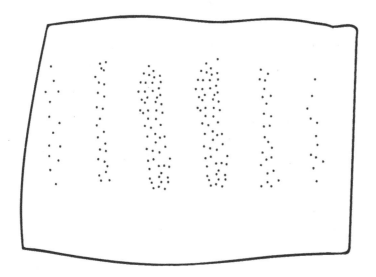

That should not be possible because each particle is being fired individually. The zebra pattern requires a photon going through one slit to mix with another photon coming from the other slit. If each photon is fired individually, it should not have anything to mix with. How are the photons creating an interference pattern when there is nothing for them to interfere with? Are the photons somehow going through both slits at the same time?

Quantum Pants

I remember once pulling a pair of pyjama pants out of the laundry and being confused because I was apparently wearing them at the same time. For a few seconds I stood in bewilderment, believing myself to be the owner of quantum pants, capable of existing in a pants superposition of location.

It was then pointed out to me that I own two pairs of the same pants and had simply never noticed before. In my defence, I was thinking about quantum physics at the time and in quantum physics you never go for the simple explanation. The simple explanation never works.

The double-slit experiment shows that light can act like a particle at the point of being fired from an emitter but it can act like a wave when it goes through the slits.

In terms of 'classical physics' – the physics of Isaac Newton where everything behaves sensibly – particles and waves are distinct things. Quantum theory was starting to blur that boundary.

ZIGGY SAYS IT'S COMPLICATED

As Einstein was collecting his Nobel Prize in Sweden, a young Danish physicist and football enthusiast[4] named Niels Bohr was taking quantum theory and applying it to whole atoms.

Atoms are made of particles called protons, which cluster in a central nucleus, and electrons, which hum around the outside like bees swarming a nest. (NB: neutrons, which also reside in the nucleus, had not been discovered at this point.)

It was known that light coming off a glowing atom did so at values unique to that type of atom. Hot iron emits different frequencies to hot nickel, for example, and conversely, it absorbs different light colours shone upon it. Previously this had been difficult to explain because light was thought to be a smooth wavy substance but once we learned that a beam of light was sometimes made of particles with specific energies it became possible to explain its interactions with matter. Photon energies come in specific values, so it stood to reason that electron energies did so as well.

In Bohr's quantum theory of the atom, electrons are not considered to be zipping around a nucleus at random. Instead, they travel across the surface of invisible spheres at specific distances. Bohr called these spheres 'electron shells', although he obviously should have called them 'Bohrbits'.

Bohr's atom was a three-dimensional version of a solar system and it is the popular picture of an atom people still draw today. The difference between electrons and planets, however, is that planets can travel around the sun at any distance they like. Gravity exerts a force at every point in space and decreases smoothly as you recede, so any orbit distance is allowed provided you move at the correct speed to avoid being sucked in.

Quantum electron shells are different. Electrons are not allowed to take any energy at will because energy is chunked into specific values (we say it is quantised).

A low energy electron will be fixed on a shell close to the nucleus but, if it absorbs a photon, it gets a boost and can orbit on a shell further out. Given that the distances between shells are fixed, only certain energy jumps are permitted and thus only certain beams of light will interact with certain atoms.

Suppose the distance between two shells is a 20-eV jump. If an electron absorbs a 20-eV photon it can thus make the jump perfectly. But if we were to fire a 19-eV photon at the atom, nothing can happen. A 19-eV jump is not permitted so the photon will carry on right through as if the atom were not there.

This means electrons cannot exist at intermediate energy values between shells. So when an electron absorbs a photon and transitions to a higher shell it does not pass through the no-man's-land between. It apparently snaps from the inner shell to the outer one instantly, in what is called a 'quantum leap'. I am not saying the electrons teleport between shells . . . but it looks an awful lot like it.

A quantum leap is an electron disappearing from one shell and reappearing on the next, simultaneously absorbing a photon in the process (if gaining energy) or releasing one (if losing it). Ironically, in everyday use the term 'quantum leap' tends to mean a huge change, but it actually refers to the smallest change it is literally possible to make.

Bohr was not sure *why* electrons orbited at certain energies and quantum leapt between them but it explained what he wanted so he jammed a bunch of ideas together and decided not to worry about it.

In essence, Bohr made a collage out of existing physics ideas like a child stealing fabrics from their parents' linen cupboard and taping them together to form a sincere but ugly picture. And, since nobody else was doing any better, everyone just accepted it and stuck it to their refrigerators.

Physics is Never Bohring

Quantum leaping, however, does explain something else pretty crucial. Protons exert an attractive pull on their electrons. It is this pull that electrons have to overcome in the photoelectric effect we saw earlier. We call the attractive property of a particle its 'charge' and it comes in two varieties arbitrarily named positive (for protons) and negative (for electrons). Particles with the same charge repel like similar ends of a magnet, while particles possessing opposite charges attract.

Charge has been known about since the days of Benjamin Franklin and his lightning-kite experiment (a real experiment by the way, not an urban legend).[5] What charge actually is gets complicated (we'll find out in Chapter Twelve) but whether you know what causes charge or not, it raises a good question: if electrons have the opposite charge to protons and are attracted to them, how come they do not spiral towards the nucleus, shrinking the atom? Why are atoms not doomed?

Bohr's answer was that this would violate the principle of quantum energy. An electron on the lowest shell, nearest the nucleus, is on the bottom rung of the energy ladder. If it were to start drifting inward it would be taking all sorts of values that are not permitted.

The only way to lose energy once you're on the innermost shell would be to step off the ladder altogether and simply stop existing. Electrons might desperately want to move towards the nucleus but the principle of quantum energy is deeper than the law of charge attraction.

Kings of the Electron

At about the time quantum theory was beginning to germinate in Europe, the undisputed don of particle physics was the British physicist J. J. Thomson – the man who showed that electrons had a negative charge, as well as discovering them in the first place.

Today, Thomson's ashes are buried next to those of Isaac Newton and, at Cambridge University, the physics department is located on J. J. Thomson Avenue. Oh, and he was knighted. And he got a Nobel Prize. As did six of his students.

But did he invent quantum pants?

Discovering the electron and its properties was Thomson's crowning glory. He had demonstrated their existence by deflecting arcs of electricity and measuring how much the arcs weighed. Since electricity had a mass, it was evidently made from particles that did as well.

When he originally announced this discovery on 30 April 1897, several people came up to him at the end of the lecture to congratulate him on having pulled off a successful hoax.[6] Nothing could be smaller than an atom, surely?

Electrons are real though, make no mistake. Two thousand times lighter than the smallest atom, hydrogen, but real nonetheless. Thomson originally wanted to call them corpuscles in honour of Newton, and the American physicist Carl Anderson wanted to call them negatrons[7] (which we can all agree is the best name imaginable) but electron caught on instead.

Among Thomson's many notable students were Ernest Rutherford, who discovered the atomic nucleus, and Niels Bohr, who showed that electrons had to orbit this nucleus in shells.

The most revolutionary discovery made by one of Thomson's students, however, was that electrons were not particles all the time. They sometimes had wave behaviour just like photons. A discovery made by George Thomson, J. J.'s son.

George was interested in light being sometimes a particle and sometimes a wave, so he decided to see if the same thing could be achieved with electrons.

If electrons had wave properties, they were obviously very small waves to have gone undetected for so long. In order to diffract electrons through a double-slit experiment, he would therefore need a tiny double slit (smaller waves need a smaller separation between slits), which is not an easy thing to build.

To get around the problem, he obtained some celluloid film like the kind used in movie cameras because in this substance, atoms are spaced in rows at regular intervals resembling a double slit on the atomic scale, and he fired a beam of electrons through it.

Sure enough, the beam on the other side split into the zebra pattern, meaning the electrons had to be interfering with each other like waves.

(NB: electrons are actually the particles Tonomura used in his experiment from earlier, but I felt that if I announced electrons were waves at that point in the chapter it would have caused mass panic, rioting and the end of civilisation as we know it. So I lied.)

Turns out electrons, which everyone knew to be particles, could be superpositioned and diffracted like waves of light. Nobel Prize for George.

It's sort of brilliant that J. J. Thomson won a Nobel Prize in 1908 for proving electrons were particles and then his son got one in 1937 for proving they were not. I like to imagine awkward Christmas dinners at the Thomson household with J. J. and George sitting opposite each other, both wearing scowls and colourful paper hats, casually polishing their prize medals while Mrs Thomson sits uncomfortably between them. 'Anyone for plum pudding, dears?'

CHAPTER THREE

Aristocrats, Bombs and Pollen

The Duke of Duality

I remember many years ago, in the hazy days of my youth, attending a university interview and being sat in front of four distinguished scientists who asked what I knew about quantum theory. Foolishly, I had mentioned it in my application letter so they wanted to grill me and take me down an energy shell or two.

I reeled off a bunch of facts about waves and particles, trying to make it look like I knew my stuff, until one of them raised her hand to halt my waffle and asked very gently, 'So, is an electron a particle or a wave?' before sitting back to watch me flounder. I am not bitter about this experience at all, but in fairness she was asking me an unanswerable question.

The mystery of electrons and photons behaving in different ways is called 'wavevparticle duality' – a term coined by the French nobleman Louis Pierre Raymont, 7th Duke of de Broglie (Louis de Broghie for short). Louis served in the military during the First World War and insisted on getting an education afterwards in both history and physics, which he thought crucial for understanding the past and future of humanity.

By the time he was in his twenties, quantum theory was the big thing in science, so he decided to write his thesis on this central enigma. Was it possible that things in the universe were neither particle nor wave and only took these forms depending on which experiment we performed? Did electrons and photons somehow hop back and forth between the two states? Could our feeble chimp-brains even handle what nature is really doing at the quantum level?

When we think of waves, we can calculate how much energy they are carrying from their frequency (how many waves hit you per second) and their wavelength (how far apart each wave peak is).

Since we can also calculate the energy of a moving particle given its mass and velocity, de Broglie posed the question: why not set these two energies equal to each other? If we know the properties of something in particle terms we can calculate its energy and then switch our brains to think of it as a wave, the energy of which we have just calculated. Energy serves as a translator between wave physics and particle physics.

Initially this suggestion was met with scepticism. Were we seriously going to say every particle had a wavelength and every wave had a mass? Fortunately for de Broglie, Albert Einstein liked his idea a lot and began endorsing it in lectures (which never hurts).

The de Broglie approach says you can take any particle you like and calculate its 'associated wavelength'. Once you have that, you can build a double slit of the appropriate size and fire your particle at it, getting an interference pattern on the other side. While we might not be able to visualise how something can be a particle and a wave at the same time, we can certainly do calculations on it and get reliable data.

And it works too. In 1944 Ernest Wollan used de Broglie's theory to diffract neutrons, thousands of times heavier than electrons, through a crystal of table salt.[1] Protons can also be diffracted the same way, although surprisingly there are no records of who did this experiment first. With hindsight, I probably should not have claimed it was me in that university application letter.

A Bit Extreme

To learn that protons, neutrons and electrons all behave like waves is both profound and peculiar. Every object in the world is made from those particles so everything we think of as matter, including our own bodies, is wavelike as well. Your body has a wavelength and if we somehow fired you at an appropriate double slit you could be diffracted.

If you're curious, an average human being launched out of a cannon at 30 metres a second would have a de Broglie wavelength of about 0.0000000000000000000000000000003 metres. If we could somehow find a way of getting every atom in the human body to line up with a double slit that size we could genuinely get diffraction occurring. I reckon you could use your volunteer from the earlier experiment where you blasted their hand off with a cannonball. The current world record for diffracting objects bigger than a single particle is held by Sandra Eibenberger who, in 2013, managed to wave-interfere a whole molecule of $C_{284}H_{190}F_{320}S_{12}N_4$. That's 810 atoms going through both slits at the same time, superpositioning with themselves on the other side.[2] Not quite a whole person, but we have to start somewhere.

ENTER HEISENBERG

Before he was a law-breaking New Mexico chemistry teacher played by Bryan Cranston, Werner Heisenberg was one of the finest mathematicians in the world. Unlike Planck, Einstein and de Broglie, who focused on experimental results, Heisenberg was more interested in taking well-established theories and twisting them to breaking point without worrying what it meant for the lab workers.

He was notoriously ignorant of real-world physics and during his doctoral viva was asked how a simple battery worked and had no idea[3] (which I find reassuring since evidently even Heisenberg faced humiliating interviewers).

Despite this practical incompetence, his mathematical brilliance was unparalleled and in 1920 he was hired to work for Arnold Sommerfeld, one of the physicists who had helped Bohr devise his atom theory.

Sommerfeld gave him a puzzle to work on which involved the mathematics of how light splits itself, which Heisenberg cracked in under a fortnight but his solution was so complicated that Sommerfeld rejected it, assuming he could not have hit the answer so quickly. That was until a few months later when the far more established physicist Alfred Landé published the exact same idea, thus getting the credit.[4]

Soon after that experience, Heisenberg transferred to work with Niels Bohr at the Copenhagen Institute in Denmark, fast becoming the world citadel for quantum research. Perhaps he felt bummed about the way Sommerfeld could not accept his genius, or maybe he just fancied working for a Nobel Prize winner (Sommerfeld was nominated eighty-four times but never won). Either way, Heisenberg relocated and became Bohr's top student and one of his closest friends.

These were Heisenberg's golden years as he found himself surrounded by the sharpest minds in Europe, devising much of the methods and equations we still use in quantum physics today. Heisenberg and Bohr would go hiking together in the mountains, go out on the town in search of women, and spend every other waking moment discussing the peculiar behaviour of particles. It was a time when Heisenberg was truly respected and happy.

Sadly, the latter half of his life was more controversial. As Nazism spread across Europe, many scientists fled the onslaught and relocated to America. Heisenberg stayed put, however, and was recruited to help the Nazis build an atom bomb.

According to some historians, Heisenberg tried to sabotage these efforts from within because in post-war interviews he described accurately how such a bomb could be built, even though his project never succeeded. Perhaps he figured it all out but kept his mouth shut to stifle the Nazi effort.[5]

However, in 2002 a series of letters between Heisenberg and Bohr was uncovered, and they paint a murkier picture. It appears Heisenberg was perfectly comfortable working on the bomb and it only failed because he did not have a good team working for him (the good scientists were Stateside) and he was himself clueless in the lab.[6] Presumably, all the equipment was battery operated.

Nobody knows for sure what Heisenberg's ethical stance was during this time. He did get himself in hot water for promoting the work of Albert Einstein, a Jewish physicist, and only avoided serious trouble because his mum got him out of it. Mrs Heisenberg was good friends with Mrs Himmler, the mother of Heinrich Himmler, head of the SS, and when Heisenberg was in trouble she phoned her friend and effectively said, 'Tell your son to leave mine alone!'[7]

When it comes to Heisenberg's politics, I suspect he simply was not thinking very hard about the ethical implications of what he was doing and just wanted to solve physics problems people gave him.

Judging him as good or evil in retrospect is difficult because we do not have much to go on, and let us not forget that the Allied side was hardly composed of saints. Robert Oppenheimer, Heisenberg's opposite number in America, was a man who once tried to poison his doctoral supervisor in true fairy-tale fashion by covering an apple in toxic chemicals – what we'd probably call attempted murder.[8]

But if we leave aside the political overtones of his later life, Heisenberg's contribution to quantum theory is still invaluable.

ELIMINATE ALL DISTRACTIONS

One summer, while suffering from a severe bout of hay fever, Heisenberg decided to spend a holiday on the island of Heligoland, which has no pollen-producing plants. During this break he hit upon a new approach to the mathematics of quantum theory, which included the origin of the thing for which he is best known today: the Heisenberg uncertainty principle.[9]

Particles have clearly defined properties we can measure. Things such as location, velocity, mass and so on. If we know everything about the initial state of a particle we can, in theory, predict everything it is going to do in the next moment. And the next. And the next.

This philosophy of determinism began with Isaac Newton and is what made physics so important. Whereas mystics of old claimed you needed to slaughter virgins and drink dove's blood to predict the future, Newton showed you could do it with a few equations and 100 per cent accuracy.

But de Broglie and co. discovered that all particles have wave properties too. Asking 'where' a particle is and 'how fast' it is going are separate questions in Newton's physics but because waves are, by definition, in motion constantly and their location is spread across a region, concepts such as speed and location are no longer independent of each other.

If you know something about the location of a wave this also contains information about its velocity and vice versa – the two properties are linked. So Heisenberg applied a similar idea to particles. Their motion and location could not really be treated as separate because they were only partly particulate.

If we have a particle and we have not yet taken any measurements on it we can specify what momentum and location it probably has with a graph such as this:

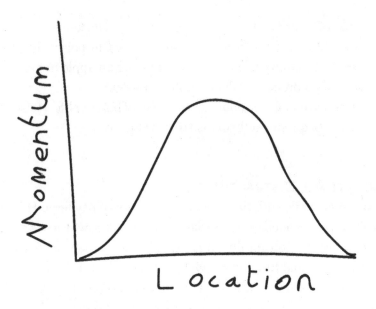

All we know is that the particle's physical characteristics will be some-where inside that hump. In normal, classical terms, when we take a measurement on a particle we are squashing that hump into a dot some-where on the grid, which pinpoints both of the values and tells us where the particle is and what momentum it has.

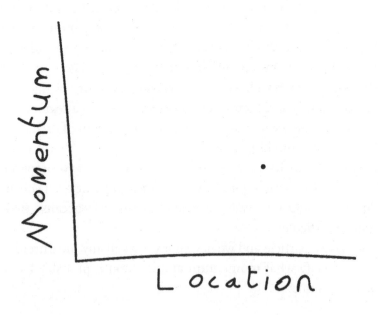

In this diagram we read along the horizontal axis to find the particle's location and then read vertically to discover its momentum. Simple enough.

But Heisenberg knew waves were different. What happens when you try to pinpoint a wave is you get a spike such as this:

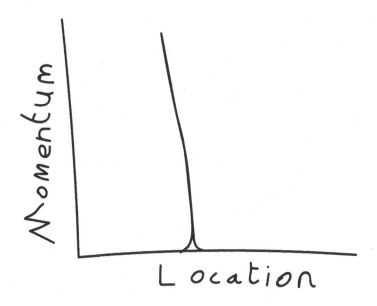

We know exactly where the particle is because we've narrowed its location to a single value on the horizontal axis, but if you read vertically you find a whole variety of momenta occurring at the same time. Because the wave's position and momentum are not separated like they would be according to everyday classical physics, if you know the location of your particle you lose all certainty about its momentum.

Or if we imagine squeezing things in the other direction we might be able to measure a precise value for a particle's momentum, but we end up with the particle existing in lots of locations at once:

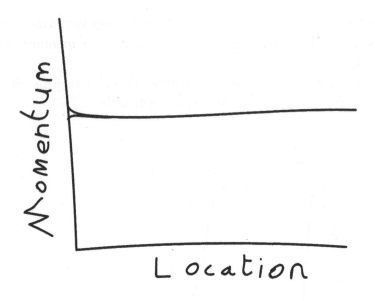

Momentum and location are linked properties in quantum theory, so we can either know where a particle is or what its momentum is, but never both at the same time.

ARE WE QUITE CERTAIN OF THAT?

This was the first 'uncertainty relationship' and the most oft-quoted one: you can never know a particle's location and momentum at the same time and if you know one of them you lose any knowledge of the other. As quantum theory evolved we discovered other pairs of properties that are linked together (we shall meet some later) but Heisenberg's original still shook the foundations of physics.

Putting it bluntly, quantum theory says it is impossible to know everything about an object because there will always be something we cannot measure. If you know enough about one property, you automatically lose information about something else.

Newton's view of a universe in which we know the future by knowing the present is thus slaughtered by a maths nerd with hay fever. You cannot know everything about the present, so correctly predicting the future is never going to happen. Ever.

Sometimes the Heisenberg uncertainty principle gets misrepresented by saying our equipment is not good enough to know all of a particle's properties but this is selling it short. It does not matter how well we build a detector and it does not matter how precisely we make a measurement. We can never pin down a particle's properties in full because it is not exclusively a particle. It is a wave too.

Asking an electron about its particle properties is like asking, 'Which letter is *War and Peace*?' or 'What colour is a rainbow?' It is trying to narrow something that cannot be narrowed.

As Heisenberg himself put it: 'We cannot know, as a matter of principle, the present in all its details.'[10] Quantum theory forces us to give up Newton's dream of predicting the future with accuracy.

In the everyday world it is easy to know where an object is and how fast it is going. That forms the basis of every sport, in fact. A game of quantum basketball would be totally unfeasible (although hilarious) because if we threw a ball with a known momentum we could no longer be sure where it was. It would blur into a ballish-cloud mid-air and, although we could watch how fast this cloud moved, we could not specify exactly where we would need to stand to catch it. The only reason we do not notice the uncertainty principle in our everyday lives is because we are very big compared to the size of the 'uncertainty clouds'. It does not cause an issue when we are doing everyday tasks but if we want to take measurements of a single particle we hit a wall of ignorance.

It also means a particle can never be brought to rest. If an electron were to stop moving around a nucleus it would have a clearly defined position and a single momentum (zero). It would be behaving as a particle only and its wave character would vanish. That does not happen and thus particles have to be in perpetual motion.

In Heisenberg's own words:

It was almost three o'clock in the morning before the final result of my computations lay before me ... At first, I was deeply alarmed. I had the feeling that, through the surface of atomic phenomena, I was looking at a strangely beautiful interior, and felt almost giddy at the thought that I now had to probe this wealth of mathematical structures nature had so generously spread out before me. I was far too excited to sleep, and so, as a new day dawned, I made for the southern tip of the island, where I had been longing to climb a rock jutting out into the sea. I now did so without too much trouble, and waited for the sun to rise.[11]

CHAPTER FOUR

Taming the Beast

Oh, Doctor Schrödinger!

Bohr's theory of the atom with its electron shells and quantum leaps was still not sitting well in everyone's stomach. It did not explain *why* only certain energy shells were allowed and the data it predicted did not always match experiment.

He had mashed his atom together from different theories like a child playing with dolls, forcing them to kiss, and he knew it would not be a permanent fix. There had to be a neater, more elegant theory nobody was thinking of. The person who cracked it was Erwin Schrödinger, a man whose personal life was the talk of the town even more than his theories.

Schrödinger was a socially eccentric, spiritually liberal, bow-tie-wearing genius who wrote prolifically on many topics including science, the arts and philosophy over a career spanning fifty years. He was an outcast in polite society and caused a scandal for living in a three-way relationship with his wife and mistress, Anny and Hilde, as well as fathering children from at least two other women (contrary to popular myth, science nerds can have very active personal lives and do not have trouble finding people with whom to be active).[1]

As interesting as he was though, the hanky-panky of Schrödinger's personal life was not the reason he got a Nobel Prize (it would have been quite the acceptance speech); it was for making a superior recipe for explaining the atom.

Humbug

Schrödinger hated Christmas. He was notoriously against religiously tinted events and so, in December 1925, decided to isolate himself from the festivities at a remote villa in Switzerland. His wife stayed home, but

43

he was accompanied in his hibernation by 'an old girlfriend from Vienna' the identity of whom has been lost.[2] Although he stopped recording in his diary during this time – so nobody really knows what he got up to – we do know he brought a physics problem in his luggage to work on over the holidays.

Schrödinger initially had little interest in quantum theory. He was a talented physicist but his area of expertise was in the behaviour of waves, not particles. Electrons, photons and protons frankly bored him. That was, of course, until people realised we needed to think about these particles as waves some of the time.

Schrödinger knew there were dozens of equations to predict the behaviour of electrons as if they were particles, but nobody had invented a method to describe them as waves. He later wrote of this time (paraphrased from German), 'this extreme idea may be wrong . . . on the other hand, the opposite point of view, which neglects waves, has led to such difficulties that it seems desirable to lay an exaggerated stress on the opposite approach'.[3]

Everyone else was using particle physics to describe atoms, but if he could come up with a wave equation to do the same thing, maybe it would give new insights. Essentially Schrödinger was trying to be awkwardly different. Given his success with the opposite sex and the Nobel Prize sitting casually on his mantelpiece, however, maybe he was onto something.

By his own admission, Schrödinger did not know enough mathematics to come up with a new law,[4] but he obviously surprised himself over that mysterious winter because as the new year dawned, he emerged with exactly what everyone had been searching for: an equation that accurately predicted the energies of electrons going around the nucleus.

Schrödinger put aside the question of how wave–particle duality worked and just focused on the wave side of things. He imagined stretching each electron across the surface of an atom like a piece of butter spread over toast, and this electron membrane could wrap around the nucleus and vibrate at certain frequencies.

Given a few input values such as mass or pull from the nucleus, Schrödinger's equation accurately predicts, using something he called a 'wavefunction', the shape an atom's electrons ought to vibrate in three dimensionally.

A wavefunction is an equation you solve to generate a list of properties an electron has at a particular point in space or time: the height of the wave, its wavelength, the speed at which it is rippling and so on.

Schrödinger's equation does a calculation on this wavefunction (it's an equation within an equation) and predicts how the electron's properties and behaviour evolve with time. Not only that, it finally explained why only certain energy shell values were allowed.

WHERE WE'RE GOING WE DON'T NEED PARTICLES

Because every electron in an atom is trapped by the nucleus, there are restrictions on its behaviour. For example, a wave can only exist in whole numbers of ripples, shown in the two images below. The left image is showing a single wave and the right is showing a double wave.

You cannot have three-quarters of a wave because it would not fit on the line. Only specific shapes are permitted.

These permitted waves are called 'harmonics' and that name, which sounds vaguely musical, is no coincidence. The shapes in the diagrams above correspond to the notes you can play on a stringed instrument, with each shape of the wave producing a different sound in the air.

When you pluck a guitar or banjo string, it vibrates with a very specific energy. A different note is a different harmonic (permitted wave) and Schrödinger's equation shows that electrons bound to a nucleus are equally musical.

We have to go into higher dimensions, of course, because atoms are not straight lines, but if you can calculate the harmonics of a three-dimensional vibration you should get the shapes electron waves can take. It's rather hard to imagine how you might actually calculate this (it was beyond Schrödinger's talents) but fortunately other mathematicians soon calculated the answers.

The first electron harmonic is spherical (shown on the left below). The second harmonic looks like two balloons squashed against each other, one at the front and one at the back (shown on the right):

These spherical and dumb-bell shapes are telling us where we are going to find an electron wave. Rather than going around the nucleus like little pellets, we have to think of electrons as vibrating surfaces taking on bulbous shapes with the nucleus at the centre. And as we go up in energy, more complicated shapes are adopted (too fiddly for me to draw ... I just about exhausted my artistic limitations with those sphere sketches but, if you're curious, Google the term s p d f shape').

These regions around a nucleus where the electrons vibrate are no longer thought of as orbits, so we call them 'orbitals' instead. Schrödingles sounds better of course.

What this does is finally explain where energy quantisation comes from. Only certain electron harmonics will fit around a nucleus so only certain energy values are possible for a given atom. In Bohr's shell theory, energy levels were something you had to stick in out of nowhere. In Schrödinger's more advanced wavefunction theory, you get the energy levels as a prediction.

Better still, Schrödinger's equation tells us these orbitals only bond together at certain angles, a prediction confirmed by the whole of chemistry.

The Schrödinger equation also gets rid of those pesky quantum leaps too. What happens when an electron moves from an inner orbital to an outer one is like a vibrating skipping rope changing wavelength as your hand changes speed. The transition looks ugly but it is a smooth process rather than an instant one. The photon emitted or absorbed when an electron changes from one orbital to another is the result of an electron wave vibrating to a new shape.

THE ONLY CATCH

Schrödinger's equation performs calculations on a wavefunction and predicted how it would change. The wavefunction itself then provides a full description of whatever we want to know about our electron and it was without doubt a triumph of mathematical beauty. Provided you did not ask what any of it meant.

All you needed to do was plug in relevant numbers, crank the handle and symbols on the page spat out reliable data, telling you what happens to an electron in a given circumstance. But what . . . exactly . . . is an electron wave?

More worrying still, the Schrödinger equation did not give the correct answers unless the sums included something called an 'imaginary number' in the sum. I know we are taking a non-mathematical approach to quantum physics in this book, but imaginary numbers are important to the story and we cannot easily bypass them. So buckle up, baby, we're gonna get math-tastic!

USE YOUR IMAGINATION

Here's the deal. If you take a negative number such as minus two and double its size, you get minus four as an answer. We can write that sum as: $-2 \times 2 = -4$.

If we multiply minus two by minus two itself, however, we get the opposite. A minus number multiplied by another minus number turns out to be a positive because we're minusing a minus, i.e. turning a minus number inside out. Two negatives multiplied together yield a positive answer, which we write as: $-2 \times -2 = 4$.

Everyone knows the square root of four is two (if not . . . spoiler alert!) but this is not the whole answer. The square root of four is actually two *and* minus two, since they are both multiplied by themselves to get four.

That means the square root of minus four is not minus two because minus two does not square itself to give minus four. So where are the square roots of minus numbers? There does not seem to be any number you can multiply by itself to give a negative number as its answer.

In order to solve this paradox, the Greek mathematician Heron of Alexandria invented a new type of number that lives at right angles to the number scale we are used to picturing. These are numbers defined as the square roots of negatives and René Descartes referred to them as 'imaginary' because they seem pretty unrealistic.[5]

We represent them with a letter i, defined as the square root of -1. The number $i2$ is the root of -4, $i3$ is the square root of -9, $i4$ is the square root of -16 and so on.

Imaginary numbers look like cheating to some people, but then again, mathematicians often invent concepts that do not appear to make sense until science invents a use for them. After all, there was a time when negative numbers would have sounded silly. Can you hold negative five objects?

Negatives are not 'real' in the sense we can count them out in our hand, but they are definitely useful to have. The charges on electrons and protons cancel to zero, so positive and negative numbers are a good system to use.

Similarly, electron wavefunctions only work if you include imaginary numbers as part of the equation. If Schrödinger's method works (and it does) then an electron's properties are vibrating not just in three dimensions around the nucleus, but in an imaginary dimension as well. What the hell, nature?

Born Free

The first man to try and give meaning to what an electron wavefunction truly represented was the German physicist Max Born. Born was intrigued by the randomness of quantum physics, which was a direct consequence of Heisenberg's uncertainty principle.

When we take a measurement on a particle, we end up discovering its properties such as location, momentum, etc. (within Heisenberg limits), but what is really peculiar is that because these properties are sort of fuzzy prior to measurement, repeating the measurement over and over can give different results each time.

If you repeat a classical (normal) experiment over and over you get the same result. Roll a ball down a ramp and you can comfortably predict where it will arrive. As far as someone such as Isaac Newton was concerned, there is no such thing as true randomness or chance. Just predictable laws of physics.

Even the toss of a coin is not random to a classical physicist. The impulse applied from your thumb, the angle of the coin as it arcs through the air and its interaction with the ground all predict how it will finally land.

If you have a powerful enough computer and feed it all the data, you can predict the outcome of a coin flip with perfect accuracy. The only reason we treat coin flips as random is because we cannot do such intensive calculations on the spur of the moment. But quantum physics is different. Quantum physics appears to have genuine randomness in its outcomes.

You may have heard the proverb that 'insanity is repeating the same mistake and expecting different results', often misattributed to Einstein (it actually comes from a 1981 pamphlet printed by Narcotics Anonymous).[6] It makes sense though. How crazy would you have to be to repeat the same thing over and over expecting a different thing to happen? As crazy as a quantum physicist it turns out.

Take the double-slit experiment. At the start you fire a bunch of electrons/protons/whatever towards a two-slit junction and end up with a zebra pattern on your detector screen. But you cannot tell, ahead of time, which stripe a particle is going to land in when it arrives at the screen – all you can do is make a guess based on probability.

There could be a 40 per cent chance of your particle arriving in the central band, then a 20 per cent chance of it arriving in the bands either side, a 10 per cent chance of landing in the next two, and so on (remember those numbers for a moment).

Effectively, when you roll an electron down a track its destination can change on each repeat. Heisenberg's uncertainty principle forces us to abandon the idea of predictable futures and accept that things happen based on probability, at the whim of some elegant although deranged quantum goddess. A particle's location is not precise until we actually measure it (it is uncertain) and when we do the measurement we can only predict a probable location, not an absolute one.

Born decided, therefore, to calculate the Schrödinger answers for an electron wave as it goes through a double slit, and discovered that the wave 'amplitude' (how high the wave is) corresponded to familiar numbers.

The peak in the centre of the detector screen (where the wave amplitude is highest) winds up with a value of 6.32. The next two bands either side have an intensity of 4.47, then the next are 3.16. These numbers do not seem to follow a pattern but they do: they are the square roots of the probability percentages we saw a moment ago – 40 per cent, 20 per cent, 10 per cent and so on.

If we do the Schrödinger calculation on an electron wave, then multiply the answers by themselves (square them), they match the probable locations for where particles are located in an experiment.

It would appear that Schrödinger's wavefunction is calculating the square root of an electron's probable location. So yeah. Gee. Thanks, Born. That makes it perfectly clear. What was I confused about?

What Does that Even Mean?

Born's interpretation of the Schrödinger equation seems to imply that probability – a human invention for saying which horse might win a race – is an inherent law of the universe that particles have to obey.

Particles are definitely particles but their locations are determined by waves of probability itself, constantly in flux. In this view of quantum theory we have to abandon the idea of anything having a definite location and say that locations are determined at random, based on laws of probability.

Areas where the Schrödinger wave is peaking are where a particle is likely to be detected, and where the Schrödinger wave is dipping are where a particle is less likely to be. Electrons, protons and photons are not really waves themselves, but their probable locations are.

We can never predict exactly where a particle is at any given time, but by using the Schrödinger equation (and assuming particle locations vibrate in an imaginary direction as well as real ones) we can calculate a probable outcome to our measurement.

Particles can therefore be pushed through the same experiment over and over but finish in a different place because their fate is never set in stone. Every cause has many potential effects and the quantum goddess is choosing which ones will happen entirely at random. Sometimes an electron will be on one side of an atom, but as its location vibrates through space it can find itself existing on the other side.

If this is all getting a bit heavy and confusing then here's a bit of fun trivia to lighten the mood: Max Born's granddaughter was Olivia Newton-John, the singer and actress who starred opposite John Travolta in the 1978 movie *Grease*.

Sadly *Grease* is light on references to quantum physics, unless we decide 'chills' is a metaphor for 'wavefunction' and when our chills are multiplying, this means multiplying the wavefunction by itself (squaring it) to get the probability of the electron's final resting place.

If so, it would be accurate to say that once we solve the Schrödinger equation we are faced with a philosophical crisis in which we have to accept the outcome of an experiment is truly random and we are thus 'losing control'.

Tunnel to Victory

Born's interpretation of the wavefunction is bold, but there is a way to test it. We can throw a particle at a wall and see if it sticks.

If you imagine a classical (ordinary) object such as a tennis ball hurtling towards a barrier, there's no question about what is going to happen – it will collide with the wall, stick there for a moment and then bounce back.

Born's interpretation for quantum particles suggests otherwise. A particle's location can be described as a constantly vibrating wave of probability.

If we throw our electron at a wall we have to consider it as a wave. Each peak on the wave means 'particle is likely to exist here' and each dip means 'particle unlikely to be here'. So if our wave approaches a wall, some of its peaks might be on the other side of the wall, like so:

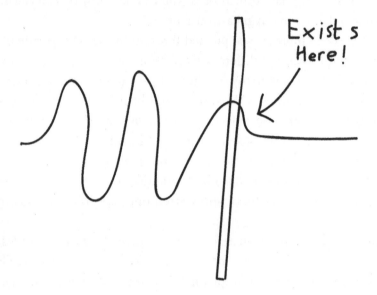

When the particle reaches the wall, as in the diagram above, most of its location stays on one side (the peaks of the wave are mostly on the left) but a tiny fraction crops up on the other side every once in a while. This is not a common occurrence because the wavefunction's value is pretty low on that side (it is only a tiny bump), but occasionally an electron should appear on the far side of a wall by chance.

This effect, known as quantum tunnelling, has been observed and documented many times, exactly in accordance with Born's prediction. There is even a device in electronics called a Josephson junction, which consists of two conducting materials sandwiched either side of an insulator. An electron would normally be blocked by such a barrier but thanks to the tunnelling effect it is possible to transport electrons through the barrier and control the flow of current by altering its thickness.

It takes a hundred billionths of a second for tunnelling to occur[7] and it is similar in essence to the double-slit experiment. A particle has a choice to bounce off the wall or tunnel through, just like it has a choice about which slit to go through. The only difference is that in the double-slit experiment the probability for each slit is 50 per cent likely but in the tunnelling scenario above, the probability of moving through the barrier is very low so we do not see it happen very often.

Tunnelling also explains where radioactivity comes from. An atomic nucleus can sometimes spit out a couple of protons and neutrons at random (called alpha radiation). This would not be possible classically since the protons and neutrons would never make it past the other protons and neutrons encasing them in the nucleus.

But now we can describe all particle locations with a wavefunction and a small part of it can dangle outside the atom. Every once in a while, at random, we should see particles tunnelling their way to the edge of an atom and magically appearing outside, which is precisely what we observe in radioactive emission.

Choose Your Words

Prior to 1926 there was a bunch of loosely connected strands to quantum theory and Schrödinger was the guy who braided it all together. He showed that wave–particle duality was linked to the energy levels of electron shells . . . which explained the shapes of atomic orbitals and all of chemistry . . . allowing us to make predictions about particles via probability.

In some circles it is loosely agreed that the early generation of quantum physicists such as Planck, Einstein, de Broglie and Bohr were working on 'quantum theory' whereas the more sophisticated stuff of Schrödinger, Born and Heisenberg was 'quantum mechanics'.

Most people are not picky about this distinction and quantum mechanics usually serves as an umbrella term for both pre- and post-1926 physics. But, for the purists, quantum theory started with Planck and quantum mechanics began with Schrödinger.

CHAPTER FIVE

Things Get Even Weirder . . . Again

Everything We Know Is Wrong

You may have noticed by now that every theory in quantum mechanics soon turned out to be incorrect. For people not familiar with science this can seem disconcerting, as if scientists are in a constant state of uncertainty (can I interest anyone in a Heisenberg pun?) but this is the normal state of affairs.

Scientists work best by taking an idea and pushing it to the limit to see when it no longer works because nothing is above question and no fact is sacrosanct. It is always better to have confidence in an idea than certainty because then it is easier to admit you might be wrong. The Schrödinger equation, for all its brilliance, is no different.

When he published it, Schrödinger openly ignored the electron's charge because it stays constant and does not need to be adjusted for. This does mean, however, that his equation dissolves if you bring your electron near a magnet.

Magnetism and electric charge have a strong influence on each other. A moving magnet can coax electrons in a wire to flow, and an electric current going in a circle will blossom a magnetic field around itself. An equation that describes the electron but ignores magnetic effects is therefore incomplete.

You Spin Me Right Round

We will explore the link between electric charge and magnetism in Chapter Eleven, but what it comes down to for now is that electrons have magnetic fields around them like tiny bar magnets, one end being north and the other being south. I like to think of them as having little harpoons skewered through their middle indicating which way their magnetic field points:

A magnetic field can be generated by spinning an electric current around in a loop of wire, so the assumption was made that the magnetic property of an individual electron must arise in a similar way. Electrons must be spinning constantly like a gyroscope to generate magnetic poles.

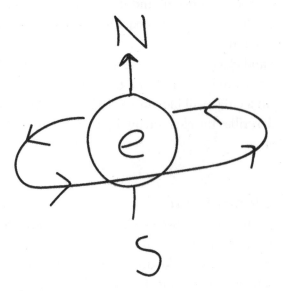

Unfortunately, when we calculate how big the electron's radius would have to be to generate a magnetic field of the right size, we get an answer bigger than a whole atom, so trying to explain an electron's magnetic field as the result of it spinning on an axis is clearly wrong.

An electron's magnetism must be due to some mysterious property they possess in addition to charge, but by the time we realised this everyone was already imagining electrons rotating, so we stuck with the name and called this mysterious property, rather unhelpfully, 'electron spin'. It does not literally mean an electron is spinning on an axis, it is just the word we use to refer to the electron's magnetic nature.

In order to probe what was going on with this 'spin' property, the German duo Otto Stern and Walther Gerlach decided to measure the spin of particles by firing them through a gateway with a magnetic field stretched across it, slightly stronger at one end to engender an overall force on the particle (rather than cancelling out on both sides).

As a magnetic particle was fired through this gateway and passed through the magnetic field, it should have been deflected at an angle, depending on how much it happened to be 'spinning' at that moment. Again, we knew it was not literally the particle whirling round, but whatever spin was it should come at lots of different values with particles deflecting all over the place.

What Stern and Gerlach found instead was that the particles flew in two directions only. Spin (whatever it is) is always the same size but either goes in the same direction as a magnetic field or against it. There were no in between values which means spin, just like energy, is another quantised property.[1]

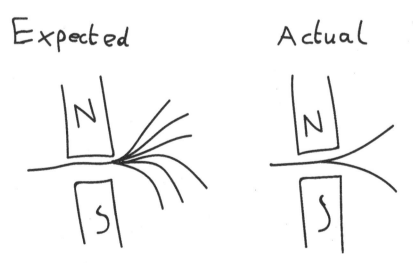

Expected   Actual

At this point, it would have made sense to stop using the word spin and refer to it as magnetic charge or something. But everyone clung stubbornly to spin and, just to be super-extra helpful, rather than calling the two types clockwise and anticlockwise, Stern and Gerlach called them up-spin and down-spin (see Appendix I for a slightly more detailed discussion of spin and magnetism).

The Schrödinger equation was then duly modified by the physicist Wolfgang Pauli to take spin into account, giving us the Pauli equation.

Pauli had a reputation for lab incompetence even worse than Heisenberg's, so much so that his colleagues referred to the spontaneous malfunctioning of equipment when he entered a room as 'the Pauli effect'. But, like Heisenberg, he was a theoretical physicist par excellence and his equation was a fantastic modification.

Schrödinger's equation tells us the probable location of a particle when we measure it and Pauli's version tells us its probable spin. Our definition of 'wavefunction' should really be expanded to mean a list of *all* a particle's probable properties including spin, energy, momentum, location, favourite film and whatever else we want to know about it.

ELECTRON'S CHOICE

So, we've split our electrons into two beams based on their spin using a Stern–Gerlach gateway and got them emerging in a fifty/fifty ratio. Some electrons have an up-spin property and some have a down-spin. Now, imagine taking just the beam of up-electrons and sending it through a second gateway. For fun.

Every electron in this beam carries on through the gateway in a straight line, just as expected. We're sending up-spin electrons through, so naturally they all come out in the upward direction only. But now suppose we rotate the second gateway by ninety degrees. The beam of up-electrons splits in two, this time flying left or right.

Spin is apparently determined along whichever axis we are looking at, so obviously if we set our magnets horizontally we should get left and right beams emerging.

We have now separated all our electrons into up or down categories and then subdivided the up-electrons into left and right.

Let's now take the up-left electron beam and put it through yet another gateway, this time set vertically like the first one. We have got a stream of up-left electrons going through a vertical gate so we should clearly get a single beam of 'up' electrons coming out the other side. We do not: 50 per cent of the electrons go down.

NEVER LET THEM GET YOU DOWN

Imagine we had a group of people and we sent them through a hospital clinic for blood testing. First, we send them through a triage that separates A-type from B-type blood. Then we take the A-type people and send them through a second clinic to distinguish A+ from A–. Once that's done we take the A+ people and send them back through the first clinic, only to discover that some of them have spontaneously transformed their blood into B-type. This is what electrons are apparently doing with their spin. But how can this be so?

Well, we assumed that when we sent our up-electrons through the left/right gateway they came out with their up-ness intact. But that would not explain what we actually see. Technically, when we fire up-electrons through a left/right gateway, all we know for definite on the other side is their left/right spin. We do not know what their up/down spin is any more, so there is no reason to assume it is still there. It looks as though measuring the left/right spin of an electron makes it forget about its up/down spin.

This presents us with a new Heisenberg uncertainty relationship: we can never know two axes of spin for a particle simultaneously. The vertical and horizontal spin can be measured individually but measuring one erases the other, just as measuring a particle's location erases its momentum. But there is something even stranger going on.

A particle's properties get erased if we are measuring its complementary property, which would imply that the very act of measuring itself somehow causes a particle to adopt properties in the first place.

The Stern–Gerlach experiment shows that the spin property we are *not* measuring gets lost and we can only get it back by measuring it again. Apparently, a particle's properties are literally not there when we are not looking. Which means Empedocles's ancient idea that our eyes are what make an object have an appearance may contain a shaving of truth after all.

ARE YOU WATCHING CLOSELY?

Looking back at the double-slit experiment, we have to wonder if we understood it correctly. We assumed the particle had a definite location when it went through the double slit but maybe it did not. Maybe the double-slit result arises because the electron is not an electron at all and exists in some freaky-deaky ghost state, only becoming a particle with clearly defined properties when we measure it at the detector screen.

Schrödinger's wave equation lets us calculate the probability of a particular outcome, but until we actually measure for that outcome, nothing is decided. The particle does not go through either slit because it is not a particle with any properties yet.

What we really ought to do is put some kind of tiny camera next to each slit and watch what the particle does at the moment of decision. If the particle is really there we should be able to see it making its choice. Seems sort of obvious now that we say it.

All right, confession. The experiment concerned does not really put a small camera beside the slits but getting into the specifics of slit measurement is quite tedious, so let's just imagine for now that it really is done with miniature camera crews.

Prior to Stern and Gerlach's results we would have expected to record footage of a particle going through both slits at once. What actually happens is one of the most peculiar results in physics. With a camera filming the slits, electrons suddenly go through one at a time like classic particles and the quantum zebra-stripes vanish.

Putting a camera on the slits is a form of measurement and that somehow causes a particle to become a particle. When we do not film them, particles do wave stuff but the instant we switch on a camera they go back to behaving normally. We expect that sort of behaviour from politicians but not from particles. I mean how in the name of Max

Planck's light-bulb-munching ghost does a particle know if our camera is switched on or not?

This is known as the 'measurement problem' for self-evident reasons. When we are not observing a particle its momentum, energy, spin and even location are wobbling about in some non-decided nothingness we never see directly. Measuring them somehow forces them to adopt properties in the real world. Particles apparently care if we're paying attention.

CHAPTER SIX

The Box and the Pussycat

The Danish Way

As the observations poured in and the equations solidified, two camps began to emerge, each with its own view on how to move forward with the weirdness.

In one camp were the philosophers who wanted to understand the true nature of quantum mechanics. People such as Einstein, Schrödinger and de Broglie.

In the other camp were the number crunchers who wanted to use quantum mechanics without worrying about what it all meant, the most prominent of which were Bohr and Heisenberg, working out of Copenhagen.

In 1930 Heisenberg wrote a book called *The Physical Principles of the Quantum Theory* in which he summarised the view he and Bohr had been cooking up over the years. He referred to their vision as '*das Copenhagen giest*', which literally translates to 'the Copenhagen ghost', although a word such as soul or spirit is more in line with his meaning. I'm afraid quantum mechanics does not quite prove ghosts are real, although it does make a pretty good case for the undead (stay tuned).

The 'Copenhagen interpretation' of quantum mechanics is a pragmatic perspective which says the deep laws of physics are too removed from human experience so we will never understand them.

It argues that just as colours can mix to form white, properties of a particle can mix to form a 'superposition state' in which it spins neither up nor down and exists neither here nor there. It is everything and nothing at the same time.

In Heisenberg's own words, 'particles themselves are not real; they form a world of potentialities or possibilities rather than one of things or facts'.[1]

Particles do not care whether they make sense to us and are not going to bend to our human limits, so we have no choice but to take them at face value. You either fall in love with what you see or run away screaming. Wavefunctions ripple in a partly real, partly imaginary way, measurement changes them and there is nothing more to say on the matter. End of discussion.

SUMMARISED

Annoyingly, Bohr and Heisenberg never specified all the minute details of their belief in full. You could sometimes pin them down to specific answers but, like the quantum particles themselves, they kept things nebulous if they could help it. Here, however, is a rough summary of what they believed.

When not being measured, nature exists in a plural of states (or an absence of them) called a superposition, but when we take a measurement we swat this superposition down like a fly against a wall, forcing it to become one thing.

Schrödinger's equation tells us the probable outcome of the measurement but nothing is decided until it actually happens and we say the superposition 'collapses' into one state like a bubble being popped. What remains is the ordinary version of a particle, sometimes called the particle's eigenstate. This is what we work with in classical physics, but before we take the measurement we have to use probability waves instead.

It is also impossible to tell which eigenstate a particle's wavefunction will collapse into, because the universe herself has not decided. We just take the measurement and see what happens. That's how Bohr and Heisenberg slept at night.

The Copenhagen interpretation is, to a lot of people, a dirty, stinking cheat. It explains nothing and says you just have to be satisfied with ignorance. Critics have nicknamed it the 'shut up and calculate' interpretation[2] because it sidesteps intuition and flatly tells you to use the equations blindly, which does not feel right somehow. The questions that arise from the Copenhagen interpretation are pretty deep too:

1) How can a particle exist in multiple states or no states?
2) Why does measurement trap a particle in one eigenstate?
3) Why is this outcome random?
4) Why can we not know all the properties at the same time?
5) Why does the everyday world follow simple classical laws if its particles do not?

The Copenhagen interpretation answers all five by shrugging its shoulders and saying, 'That's the way it is, pardner.' Which did not work for some people, least of all Einstein.

ENOUGH IS ENOUGH!

Einstein and Bohr had a complicated bromance. They respected one another's intellect, but disagreed vehemently about quantum mechanics. At every conference they attended an argument would ignite and they wrote scores of letters privately and publicly about how mistaken the other was. Had Twitter been invented, their verbal sparring would have rivalled such infamous feuds as Selena Gomez vs Justin Bieber or Kanye West vs the entire internet.

In one famous letter to Max Born, Einstein said he could not accept a theory that made nature random and finished it by saying, 'The theory says a lot but does not really bring us any closer to the secret of the old one. I, at any rate, am convinced he does not throw dice.'[3] The 'old one' referred to here was Einstein's codeword for an impersonal God.

According to Heisenberg, who was good friends with both men, when Bohr heard of Einstein insisting God was not playing dice he responded with, 'It is not our business to prescribe to God how He should run the world.'[4] #Burn.

THE MOON, BEAUTIFUL

Einstein's chief objection to the Copenhagen interpretation was that the purpose of physics is to find out how things work. Nature was telling us something through quantum mechanics, and we had to figure out what. As he saw it, Bohr was giving up just when things were getting interesting.

In one heated exchange with the Copenhagen supporter Abraham Pais, Einstein picked on the silliness of the measurement problem and asked if Pais believed the moon stopped existing when he was not looking at it.[5] In the Copenhagen interpretation things do not have defined properties until measured, so if nobody observes the moon its wavefunction is technically in a superposition, existing in all sorts of places and states at once.

The snag is that this is an objection of gut feeling rather than evidence. Just because we like to believe the moon has existence when unobserved does not prove it does. Why shouldn't the moon disappear when nobody is measuring it? Can you prove it does not?

From Bohr's perspective, human minds evolved for the purpose of foraging for berries on the plains of Africa, not doing advanced physics. Sooner or later we were bound to run into some kind of wall. A human trying to understand quantum mechanics would be like a household thermostat trying to understand the plot of the James Bond movie *Quantum of Solace*. Come to think of it, it would be like an actual human trying to understand the plot of *Quantum of Solace*. It is simply not possible.

The picture of these debates that filtered through to the public consciousness was that Einstein was over the hill and Bohr was cautiously hailed triumphant. In some accounts Einstein is uncharitably portrayed as a once-great man no longer at the cutting edge of physics, ranting about how science was better in his day.

This probably gives people a sense of satisfaction because it is nice to know Einstein had limits, but in reality he grasped quantum mechanics perfectly. That is why he tried so hard to find a fault with it – he knew what it really was.

SCHRÖDINGER'S INFAMOUS ZOMBIE CAT

Einstein's godly dice and jumping moon were shrugged off by Bohr as emotional objections that made no difference to scientific fact. It was

Schrödinger who then decided to have a go at fighting the Copenhagen juggernaut.

I like to imagine him straightening his bow tie, rolling his sleeves up and tag-teaming Einstein out of the wrestling ring, saying something like, 'Step back, Alby, ya boi Schrödinger's got this,' supported by a small harem of cheerleaders on the sidelines. Five of them pregnant with his children.

Schrödinger decided to go a different route and attacked superposition rather than measurement. In the November 1935 edition of *Naturwissenschaft*, he published an essay that should, he argued, cripple the Copenhagen interpretation for good; it relayed what has now become known as the Schrödinger's cat paradox.

Imagine a cat sealed inside a steel chamber and left for one hour. Inside the chamber, a radioactive material is placed next to a Geiger counter, which will detect whether a particle has been emitted from it or not. It's impossible to predict whether an emission will take place by the end of an hour but we can chose a material whose wavefunction gives a probability of 50 per cent likelihood after this time.

The particle's location is 50 per cent likely to be inside the nucleus and 50 per cent likely to have tunnelled out and hit the detector. But here's where it gets interesting. The Geiger counter is rigged to a fiendish device that trips a hammer and shatters a flask of hydrocyanic acid next to the cat.

The Copenhagen interpretation says the particle takes both options, existing in a superposition of location inside the nucleus and outside it, so the hammer is simultaneously tripped and not. This means the acid flask is simultaneously shattered and left alone, and the cat is simultaneously dead and alive. If you take the Copenhagen view seriously, you have to accept something absurd. Superposition had to be wrong.

I do not know why Schrödinger did not just make it a dart or something which would execute the cat painlessly, but he chose to slowly corrode the cat over the course of several minutes via an acid bath. Schrödinger was an odd guy.

It's also not clear why he chose a cat. He definitely owned a dog named Burschie[6] and there are some claims that he owned a cat named Milton,[7] but they may be apocryphal. Perhaps he was writing his essay one morning when Milton thought it would be a good time to defecate on his desk and Schrödinger immortalised him for ever. Who knows?

Sometimes the Schrödinger's cat experiment is misrepresented as 'you do not know if the cat is alive or dead so you have to imagine it both ways'. But that is missing the point.

The Copenhagen interpretation literally says a particle can be in a superposition, which means anything interacting with it can be in a superposition too. When you open the box you are collapsing the wavefunction of the cat into a dead or alive eigenstate with 50 per cent likelihood, but until then it is both simultaneously.

This is why carrying out the experiment for real would be pointless. Schrödinger's cat would never be observed in a superposition because that's the point of Copenhagen – you measure the eigenstate a superposition randomly collapses into, never the superposition itself. Opening the box in the real world will not present you with the dead/alive superposition but a classic eigenstate of either a living kitty or a mess and a guilty conscience.

CHAPTER SEVEN

The World is an Illusion

Now You See Me, Now You Do Not

The Disney/Pixar movie *Toy Story* and its sequels are really about quantum mechanics. The main characters are toys who, when being observed by their owner Andy, behave like ordinary toys, but when he stops looking they come to life.

Andy never sees the toys in their live state and only ever observes their classical behaviour. He also never seems to notice that they shift position in between each playtime (as kind-hearted as he is, he ain't the most observant kid). But if he were to watch closely he might notice that each time he observes his toys, they are in a slightly different location.

Particles are similar. When we look away from them they seem to be doing something very different to when we are watching. We can make guesses about where they are likely to wind up by using the Schrödinger equation, but we can never predict exactly what is going to happen each time we walk into the room.

Suppose Andy began keeping a close tally of where each toy shows up and what state it is in each time. He might begin to notice that he can only describe their observed state using probability. For instance, there could be a 90 per cent probability Woody will be where Andy left him last time, but it is conceivable (the probability is not zero) that Woody is now on the other side of the room, or even outside the building somewhere.

If Andy was a Copenhagenist he would come up with a neat mathematical explanation for what was happening. His toys exist in all locations simultaneously, arranged in every possible configuration, until he enters the room, at which point their superpositions collapse into something classical. But here is the infinity and beyond question: what is it about Andy's presence in the room that triggers the toys' wavefunctions to change?

There's a scene in the first *Toy Story* where Woody explains to a group of other toys that they are going to have to break the 'rules' of being a toy, but what exactly are these rules and who decides them?

If we set up a camera in Andy's bedroom, would the toys come to life? What if they are being watched from behind and do not know about it? Why do they stay in their living state when another toy observes them? Why does the dog Buster not count? What about an artificially intelligent robot? Would a chimpanzee trigger a collapse? What about the bit where the evil child Sid sees Woody talking and has a nervous breakdown, causing him to give up on school and choose to work for the refuse collection services by *Toy Story 3*? (Keep an eye out, it's definitely Sid in a cameo.)

Trying to figure out the philosophical implications of measurement in the *Toy Story* universe is a nightmare, but that is what we have to do in quantum mechanics. The measurement problem is not just counter-intuitive, it raises serious questions about what makes reality real.

Sometimes the measurement problem is muddled up with the Heisenberg uncertainty principle but they are not quite the same. The measurement problem says a particle chooses what state to be in when observed, the uncertainty principle says that even when we make that observation, we are forced to abandon all other information about it.

An analogy for Heisenberg uncertainty would be something like Andy having an eye problem and needing specially tinted glasses to see. When he looks at his toys without these glasses he can tell what colour they are, but the image is blurred. He has certainty of colour but not shape. If he puts his glasses on, the toys become sharp but he is looking through coloured glass and cannot tell what colour they are any more. He can choose to measure colour or shape but never both.

You've Got a Friend in Me

In 1932 the physicist John von Neumann decided to break the double-slit experiment down and analyse every aspect mathematically to discover which stage was responsible for causing the collapse of the wavefunction. Some aspect of the measurement process had to be special.

He studied the particle being generated, he studied it leaving the emitter, touching the slits, moving through them, emerging on the other side and so on, calculating the relevant wavefunctions as he went. At the end, to his great annoyance, he found that none of the stages were special. Every part of a quantum mechanical experiment is physically equivalent to every other.[1]

This is pretty damaging for the Copenhagen interpretation because if nothing could be shown to cause wavefunction collapse, why is it collapsing in the first place? Bohr and Heisenberg shrugged their shoulders, Einstein scratched his ruffled hair and Schrödinger had his bedroom door closed. It's best we do not ask what was going on inside.

The most widely discussed (and poorly represented) solution to the measurement problem was eventually posed by the Hungarian physicist Eugene Wigner, who extended the Schrödinger cat experiment to include the scientist running it herself.

We have a cat sealed inside a box, existing in a dead–live superposition and when the scientist opens the box, she finds it collapsed into one eigenstate or the other. Fifty/fifty probability either way. But suppose the experiment is being done in a closed room while Wigner is waiting outside.

If we take the Copenhagen interpretation seriously then the particle both triggers and does not trigger the detector, killing the cat and not killing it at the same time. From Wigner's perspective, his scientist friend would thus open the box and simultaneously discover a dead cat and living cat. Wigner's friend would be in a superposition of relief that the cat was alive but also in a state of horror, looking for the scrubbing brush and spatula. It is only when Wigner opens the door to the lab and finds out what is going on that the wavefunction of his friend collapses.[2]

But this is preposterous. Wigner's friend could not be in a superposition because human minds have never been observed doing that. At no point do humans simultaneously observe and not observe something. Conscious awareness is something that always exists in eigenstates. Therefore, said Wigner, consciousness must be the thing collapsing wavefunctions.

According to Wigner, when we talk about measurement we are literally talking about a sentient mind making an observation and since a mind cannot exist in superposition, the particles it observes cannot either.

To be clear, Wigner was not some lightweight yahoo with an internet diploma in quantum-avocado studies. He had a Nobel Prize to his name and a reputation as a hard-nosed physicist. Introducing consciousness to the debate was not something he relished doing but he saw no alternative.

Be Cautious Here

Consciousness was and still is a mystery, but since every other part of the experimental chain can be described, wavefunction collapse had to occur inside a conscious mind, the one place we did not fully understand.

Wigner's interpretation implies a few seriously surreal things though. For starters, it would mean the mind has an effect on particles rather than the other way around.

Second, it would imply that the entire universe was in a state of super-position for billions of years until conscious creatures evolved to observe it. Surely at some point early in human history when the population was small and less spread out, there must have been moments when nobody was watching the moon. Did it jump on these occasions? If so, why did the Earth not lose its stable position in orbit around the Sun and go careening off into oblivion?

Is something always looking at the moon to stop this happening? Could there be a supreme mind consciously observing the universe at all times, keeping it in check when nobody else is observing?

To that end, here is a provocative limerick allegedly penned by the philosopher and theologian Ronald Knox on the subject of observation:

There was a young man who said 'God
Must find it exceedingly odd
To think that a tree
Should continue to be
When there's no one about in the quad.'

Reply:
'Dear Sir: Your astonishment's odd;
I am always about in the quad.
And that's why the tree
Will continue to be
Since observed by, Yours faithfully, God.'

THE MAGICAL MIND

Wigner's consciousness idea was plucky but naturally it has been misinterpreted, particularly by certain members of the New Age movement. You may not have come across so-called 'quantum spirituality' teachings but I can brief you.

When you look up quantum mechanics on the internet you will come across, along with genuine articles about the science, articles about crystals, exercise, left-wing politics, Buddhism, Hinduism, vegetarianism, yoga, self-identity and meditation. All of these are interesting topics worthy of discussion, but they are not related to quantum mechanics in any significant way.

Quantum spiritualism is where 'aspirational philosophy' gets involved because many spiritual teachers claim that since consciousness impacts reality, you can make things happen by thinking about them. I am paraphrasing their teachings a little here, but that is what they do with quantum mechanics so I think it is fair game.

I need to make it plain at this point that spirituality is an important topic, which everyone needs to address in their lives at some point. Indeed, many of the founders of quantum mechanics were deeply spiritual people (especially Schrödinger and Pauli). But we have to draw a line at what is and is not true.

Observing something might trigger wavefunction collapse but it does not determine which eigenstate we end up with. That is still random.

Each possible eigenstate has a 'probability amplitude' associated with it in the Schrödinger equation, which tells us the likelihood of it coming to fruition and the final state is determined by that, not the measurement.

While it might be acceptable to say consciousness causes an eigenstate to materialise, anything that says you can influence how it does so is immediately wrong. You are an observer of reality, but you do not influence its form in a quantum sense. If you want to make the world a better place, then I am afraid you have to go the old-fashioned way and be a good person.

Wigner's version of the Copenhagen interpretation has, in the past, been an interesting part of the debate and it solves the cat paradox nicely since the cat's consciousness is what collapses its own wavefunction. But if we are honest this is still a cheat.

It says we do not know how a wavefunction collapses or what consciousness is, but these two intangible things somehow collaborate to present us with our world. It is once again pushing the mystery into a dark cave and saying, 'This is where the answer happens, no peeking!'

The brain is mysterious but mysterious does not mean 'outside the laws of nature': it just means we do not know the details yet. There are no reasons to invoke the supernatural when accounting for quantum observation and quite a few good reasons to ignore it.

CHAPTER EIGHT

Quantum Must Die

Albert Einstein Attacks Quantum Theory!

That was the *New York Times* headline on 4 May 1935 when the greatest physicist in the world, then fifty-six, turned against his own creation like Frankenstein abandoning his monster. Einstein always had misgivings about quantum mechanics, of course, and by 1935 he was dedicating all his time to its demise, with the help of his young assistant, Nathan Rosen.

After years of blood, sweat, tears and a little help from a Russian physicist named Boris Podolsky, they finally found a chink in Bohr's armour, which has since become known as the Einstein–Podolsky–Rosen problem, or EPR for short.[1]

Does the Cat Have to Be Executed?

The Schrödinger equation describes a particle as a list of square-rooted probable properties that change over time – a wavefunction – and there is no reason we have to limit this to one particle.

A helium atom has two electrons occupying orbitals around its nucleus. Since wavefunctions can mix, the wavefunctions of both electrons can be combined to form a single 'two-electron wavefunction'. It is more complicated to handle mathematically but it presents no problem in terms of the physics.

We can go even further and add our electron-pair wavefunction to the proton and neutron wavefunctions in the nucleus, giving a wavefunction for a whole atom. In practice, doing such calculations is difficult and we usually have to take shortcuts (see Appendix II), but in theory we can calculate the wavefunction of anything we want, no matter how many particles it contains.

Let us start with two electrons whose wavefunctions we combine. Every value we calculate on that wavefunction is now dealing with the pair as a single unit, rather than the individual electrons.

A wavefunction for a pair of electrons can tell us that one of them is up-spin and the other is down-spin, but it cannot specify which is which until measurement. Because their wavefunctions are combined we have to describe the ensemble rather than the members.

Schrödinger referred to two particles linked in this way as 'entangled' since we treat their properties as knotted together and cannot predict which particle will emerge with which property when the measurement happens.[2]

We could think of an analogy by once again taking Schrödinger's box but sealing two cats inside. Classically the two cats will be distinct. Mittens is a ginger cat with shaggy fur and long whiskers, while Boots is a black cat with coarse fur and short whiskers. But in quantum mechanics all we can say is that inside the box there exists the state of 'ginger cat' and 'black cat', along with the state 'coarse hair', 'shaggy hair', 'stubby whiskers' and 'long whiskers'.

We might open the box to find one of the cats is black with shaggy fur and stubby whiskers while the other is now ginger with coarse fur and long whiskers. It is impossible to say which was originally Mittens.

Mittens does not exist any more because the features that defined her are no longer one thing, they are muddled up with Boots. So I am afraid the cat does have to cease existing, but it can get hybridised with another cat via entanglement.

Divide and Confuse

Suppose you have a particle in an unmeasured state, i.e. its spin is a superposition of up and down simultaneously. Now split it in two (do not worry how, just suppose you can) to form two daughter particles from your original. The wavefunction of the original parent particle has been splintered, but all its information is still there. The daughter particles have the combined properties of the parent but it is still not possible to say which has which.

Using classical common sense we would conclude one of the particles is up-spin and the other is down-spin. But quantum mechanics says properties of particles are not determined until measurement. The pair exists in an overall up/down state and neither has decided which way it is going to be yet.

Of course, if we now measure one of these particles it has to pick an eigenstate. Let's say it picks up. The wavefunction for the pair is still up/down overall, and since one of the particles has chosen to be up, the unmeasured one has to become down at the same instant.

The unmeasured particle is not allowed to stay in a superposition state because the overall wavefunction, according to Schrödinger, is up/down at all times. If we measure one of our particles into an eigenstate, the other has to take the corresponding eigenstate.

In order for the unmeasured particle to collapse into the down state, however, the one which *did* get measured has to let it know. A message would have to pass between them saying something like, 'I've collapsed into up-spin, you'd better collapse into down-spin now!'

It would not matter how far apart the particles were either. We could do this on opposite sides of a room or opposite sides of a planet and the outcome would be the same. Quantum mechanics predicts that an entangled pair of particles must transfer information between them in zero time and this is where the EPR paradox arises. A signal cannot go this fast because it violates the theory of special relativity.

RELATIVELY SIMPLE

Quantum mechanics was a team effort. Planck was the manager, Einstein was the captain, Bohr was goalkeeper (the actual position he played), de Broglie, Born, Sommerfeld and Pauli were midfield, then Heisenberg and Schrödinger were strikers with Stern and Gerlach in defence. Although typically Schrödinger was usually off to one side flirting with the other footballers' wives. Oh, and their mascot was a cat. Obviously.

Over thirty Nobel Prizes have been awarded for the development of quantum mechanics and while Einstein was a key figure, he was one among many. The theory of special relativity on the other hand was pretty much Einstein all the way.

There are two theories of relativity: the general theory he came up with in 1916 and the special theory in 1905. We will be talking about general relativity towards the end of the book, but the EPR paradox is all about special relativity so let's get familiar.

The theory of special relativity states two things:

1) Nobody's perspective is special when taking measurements.
2) The speed of light is the same for everyone, no matter how they are moving.

The first postulate means nobody can truly say their measurements are objective. For example, you probably think you are sitting still right now. You are not. You are orbiting the Sun and your body is now 90 kilometres away from where it was when you started reading this sentence. This means your velocity will be different depending on who is holding the speedometer. From your perspective your velocity is 0 km/s but from the Sun's point of view you are moving around it at 30 km/s. Special relativity says that both answers are valid, they are just relative to different surroundings.

Suppose you shot an arrow away from you at 20 m/s and your friend was riding a bicycle alongside it at 15 m/s (presumably she had her hand repaired from when you blasted it off with a cannonball earlier).

You would say the arrow is moving at 20 m/s but if you were to ask her, she will naturally say the arrow is travelling at 5 m/s relative to her and the bicycle. Both your numbers are true because all speeds are equally valid. Apart from the speed of light. That is 299,792,458 m/s for everyone. Everywhere. All the time.

You may have learned in school that light slows down as it moves through glass but this is a tiny bit misleading. The photons are getting absorbed and re-emitted by atoms within the glass, which delays the overall travel time, but the speed of photons in between each atom is still 299,792,458 m/s. It is a constant value of the universe.

If you shine a torch beam away from you, the beam speed is 299,792,458 m/s. Your friend is once again pedalling her bicycle parallel to the beam at, let's say, 100 m/s (she is very physically fit). If you were to ask her how fast the beam was going now, she would also say 299,792,458 m/s. The same as you. But that seems wrong. If she is moving next to the beam at 100 m/s, surely she should measure it as slower than your answer, like she did with the arrow? Special relativity says otherwise. The speed of light is the same for everyone.

Even if she was moving at 99 per cent of the speed of light, her speed-ometer would still measure the beam at 299,792,458 m/s. You would expect her to get a different answer but she gets the same every time.

Or suppose she cycles towards you, and the beam advances in her direction, head on. Should she not perceive the beam faster because she is approaching it as it approaches her? No, says Einstein. She sees the beam at the same speed.

There is an easy way to prove that the speed of light is always the same for every observer (see Appendix III) but if we take it as a given for now, the implications are truly bizarre.

If two people are moving at different speeds yet measure light at the same speed, something must be distorting between their reference frames. If they are measuring two different scenarios but getting the same answer then something is off and the culprit, said Einstein, is time itself.

If we accept the two postulates of special relativity, time must be pass-ing at different rates for two moving observers. As a person travels faster, their clock slows down and, as a result, a beam of light seems just as fast, since they are moving through time slower. Time is running slower for your friend on her bike, and therefore when she measures the speed of the beam going past it seems just as fast.

She would not sense this time dilation, however. If she were to look at her watch she would not see the second hands ticking any differently. All the particles inside her brain are experiencing time at a stretched rate, too. From her perspective, you are the one whose timeline is being warped. She sees you being sped up like a movie played on fast-forward and neither of you gets to say your time is 'correct' because all measure-ments are equally valid.

Sounds made up, right? Well, in 1971 Richard Keating and Joseph Hafele put special relativity to the test by synchronising two atomic clocks and separating them at different velocities. One was placed on board a commercial jet-plane (with a ticket made out to Mr Clock) and flown around the world eight times, while the other remained on the ground.

When the flight was over Hafele and Keating compared Mr Clock's reading with his brother and found he was lagging by exactly the amount of time Einstein predicted.

It is a small effect to be sure (Mr Clock lost a few nanoseconds only) but as you move faster you really do age a little slower. Up to a point.

The time dilation effect cannot go on forever because there is a slowest rate time can pass – it can stop altogether. Guess which speed causes that to happen? It's 299,792,458 m/s. The speed of light is really the speed at which time stops, so this is as fast as anything can ever go.

When we talk about the speed of light and say it is the fastest anything can travel, we are not phrasing it correctly. It would be better to say the faster you go, the slower time goes until you hit 299,792,458 m/s at which point time stops for you.

There is nothing special about light per se; what is special is that the universe has a maximum speed and light happens to move at it. Time distortion sets a speed limit on how fast anything can go, which makes quantum entanglement impossible.

In special relativity nothing can travel between two particles faster than 299,792,458 m/s, but information between entangled particles has to do precisely that. The particle that chooses up-spin has to send a communiqué faster than light to its partner, telling it to collapse. A phenomenon Einstein called 'spooky action at a distance'.

SPOOKY CAT

Einstein, Podolsky and Rosen highlighted this mismatch between quantum mechanics and special relativity but also offered a solution. Quantum mechanics was at fault and Einstein was correct. Big surprise there.

Entangled particles have to be deciding ahead of time which way to spin if measured. They make this agreement as they are being generated in the entangler (the casual name I've given to a device that creates entangled pairs) and then they follow their prearranged flight path.

They have a conversation that goes something like:

Electron: Hey dude, if we come across a Stern–Gerlach spin detector you should go up-spin and I'll go down-spin, OK?
Other Electron: Wait, why do I have to go up-spin?
Electron: (Sigh) You always have to be like this, don't you?
Other Electron: Like what?
Electron: Difficult.
Other Electron: I'm not being difficult, bro, I just think we should be fair about spin-states.

That wording is *exactly* how Einstein phrased it.

In Einstein's view of entanglement, the particles are not deciding at the point of detection what they are going to be, they have a predetermined answer which we are discovering. Superpositions do not really exist. Only ignorance.

Suppose we had a red cat and a green cat, which both get put into boxes and sent off to opposite ends of the solar system. If we open one of the boxes at our end we might find the red cat and thus instantly know what the other box contains. The information 'green cat' has metaphorically crossed the universe towards us but there is no violation of relativity here because nothing has to literally cover that distance.

The quantum view is that the cats have not picked properties yet and decide at random, communicating telepathically with each other at the point of measurement, faster than light. Einstein's view was that the cat's properties were always there; we just cannot see them until we measure.

After publishing this paper, Einstein and Rosen continued to work closely and enjoyed a lasting friendship. Podolsky, on the other hand, faded into moderate obscurity, although it has been claimed by some historians that he worked as a spy for the KGB during the cold war, operating under the insanely cool code name: Quantum.[3]

Oh, and by the way, I know every analogy in quantum mechanics seems to involve opening or closing a box of some sort. I will go a different way for the next one. I promise.

EINSTEIN, THE BELL TOLLS FOR THEE

Einstein's theory was ready to do some well-deserved laurel resting, but along came a Northern Irish scientist named John Stewart Bell who threw a monkey in the mechanics in the 1960s. A passionate scientist all his life, Bell disliked pomposity and went to great lengths explaining physics to the general public and, in the process of trying to find a way of describing the EPR paradox, he came up with something rather curious.

Bell's theorem, the idea which got him shortlisted for the Nobel Prize,[4] was a game-changing way of putting the EPR paradox to an experimental test. Tragically, Bell never received the Nobel Prize because he died before the committee made their decision, and the prize is never awarded posthumously. His theorem, however, lives on.

Let's entangle two particles, call them Alice and Bob, and send them both through separate Stern–Gerlach gateways on opposite sides of a room. We can have our gates aligned vertically, horizontally or diagonally for each measurement we take and, for now, let us say Einstein was right: both particles pre-determine which spin they are going to adopt upon measurement.

Vertical gateways yield up/down spin results, horizontal gateways yield left/right spin results and diagonal gateways yield ... umm ... north-east/south-west results?

The horizontal and vertical spins of a particle are independent, but the diagonal spin is not. Whatever vertical or horizontal spin a particle has will influence which way it chooses to go if measured diagonally.

Think of it like this. If we set both our detectors vertically then if Alice is up, there is a 100 per cent chance Bob will be down. If we set Bob's detector at right angles, however, we do not know what the outcome will be. Alice will be up but there is a 50 per cent chance of Bob picking left or right.

But if Bob's detector is at a forty-five-degree angle, that makes it half-way between the two extremes, meaning we can predict its spin with an accuracy halfway between 50 per cent and 100 per cent. We can predict which diagonal outcome Bob will pick 75 per cent of the time.

If a particle is up vertically, its diagonal spin is 75 per cent likely to be north-east (i.e. pointing slightly upward) and if it is down its diagonal measurement is 75 per cent likely to be south-west (pointing slightly downward).

According to Einstein, both Alice and Bob have already chosen whether they are going to be up or down, which Bell pointed out means they have a slight preference for making one diagonal decision over another.

We cannot measure an electron's diagonal and vertical spin at the same time (curse you, Heisenberg) but we *can* measure their entangled partner. If we measure the vertical spin of Alice and find it to be up, Bob should have the opposite spin pre-determined (down), which will therefore influence what it does diagonally. If Alice is up, Bob is 75 per cent likely to point south-west in a diagonal gateway.

Now say we do the experiment a hundred times. If we set the Alice detector vertically, then 75 per cent of the time she picks up, Bob will be south-west and vice versa.

If, on the other hand, Bob and Alice genuinely have not made their minds up until detection, the results of the experiment should violate the 75 per cent value and we will get a different number. This would possibly answer whether superpositions really exist.

In 1982, the French physicist Alain Aspect managed to build a real working entangler machine according to Bell's specifications and ran an EPR experiment in the hopes of validating Einstein and dooming Bohr's Copenhagen interpretation for good.[5]

Bell hoped it would reveal a 75 per cent match for Alice and Bob, proving there were hidden properties decided in advance by the particles. But the numbers were off. Completely off. Bell's 75 per cent number was not observed, which means Einstein's classical pre-determined explanation for the EPR paradox is wrong.

Particles are not deciding in advance what to be when measured. Somehow, they are truly deciding their state at the point of measurement, even when separated by distances forbidden by special relativity. Quantum weirdness prevails.

This does not necessarily prove that particles are sending messages faster than light, but in all honesty nobody is sure what it *does* prove.

Something going faster than light between particles is one explanation. Another is that the particles are linked via tiny wormholes transcending the dimensions of our universe. Another is that entangled particles are not separated at all and there is a spatial illusion that makes us humans think they are apart when really they are together.

We cannot make sense of it but somehow two entangled particles stay in communication no matter how far apart. What you do to a particle on Earth will instantly affect its twin on the moon without time for a message to pass between. Your guess for what is going on is as good as anyone else's. Personally, I'm going to say it's goblins.

CHAPTER NINE

Teleportation, Time Machines and Twirling

Instant Instagram

If we have two electrons which we entangle on Earth and send to opposite ends of space, the entanglement link allows them to communicate instantly. Could we use this to send signals faster than light then? Sadly, the answer appears to be no.

Let us suppose we have two entangled particles with undetermined spins and we seal them inside two boxes . . . wait, no, I said we would avoid boxes in the next analogy . . . we seal them inside two chickens. One chicken is sent to Mars and the other is sent to Neptune.

When our Martian colonist opens up her chicken and looks inside, the particle adopts an up state. This is interesting, she thinks. She knows immediately that the other experimenter on Neptune has a down-spin particle inside her chicken. But she has no way of telling her friend about this result, other than sending a message via ordinary channels.

There would be no way to send a signal via the entanglement link because all the link does is communicate which eigenstate a particle has adopted. We have no control over that (it is random) so we have no control over what messages get sent between entangled particles.

If our experimenter *could* somehow convince a particle to take the up state or the down, she could use a series of chickens to encode binary information, which her friend could uncover on Neptune, but this is not possible.

A superposition cannot be influenced without measuring it and measuring it collapses it, defeating the whole purpose. The only information you could ever find out through entanglement would be the results of another scientist's lab book. Faster than light communication is not on the cards. Teleportation though . . .

Beam Me Up

On 4 July 2017 a group of physicists working in China published their new world record – the longest distance teleportation ever executed.[1] The previous record had been set in 2012 when a group of researchers managed to teleport something a distance of 143 kilometres between mountains on the Canary Islands[2] but this new record beat it by a factor of ten.

The team, led by Pan Jianwei, performed a quantum teleportation from their laboratory in Tibet to the satellite Micus, orbiting 1400 km above the Earth. This is Earth-to-space instant transportation and the Trekkie in me is beginning to tingle.

Quantum teleportation was outlined in 1993 by Asher Peres, William Wootters and Charles Bennett while working at the University of Montreal.[3] We already know particles going faster than light is forbidden by special relativity but, thanks to entanglement, information about a particle's state can sneak around the problem.

Consider, once more, Alice and Bob. After being created by the entangler, they are sent to different locations we want to teleport between. Bob is sent to a satellite and Alice is kept in our laboratory on Earth. We do not know any of Alice or Bob's individual properties yet (that is what makes them entangled) but we know their overall state.

Now we introduce a third particle, the one we want to teleport. We will call it Cathy. Cathy is in a known state but if we bring her into contact with Alice, we can entangle the two of them in a superposition of their own.

We have to make sure we do not accidentally measure Alice in the process because that will collapse the entanglement bridge she has with Bob, but if we are careful (using devices called CNOT and Hadamard gates) we can get Cathy to shed her eigenstate and entangle with Alice.

Effectively, we open a room we know Alice is inside but do not look as we hold the door for Cathy to be ushered in. Cathy and Alice's state is now in a superposition so we've lost information about each one, but Cathy's original identity is in there somewhere, being shared among both.

Since we have entangled Alice with Cathy without breaking the original entanglement to Bob, the Cathy information is now shared with him also, even though he is in space. We have formed a triple entanglement scenario Schrödinger would no doubt have approved of.

If we do things just right, we can now take a measurement that tells us *some* of Cathy and Alice's information but not all of it, keeping the rest entangled with Bob still.

Suppose we do a measurement on Cathy and Alice that reveals their hair colour, but not eye colour. The hair-colour information is collapsed but eye colour is still up for grabs. If we now radio up to the satellite where Bob is and tell them to do an eye-colour measurement on Bob, we have a good chance of finding that Bob now has Cathy's eyes.

Cathy herself has not been physically transferred through space, but part of her identity has. It is as if Bob is a blank canvas and Cathy's image is peeled off and stuck to him.

Peres and Wootters wanted to call this phenomenon quantum telepheresis, but Bennett insisted 'teleportation' sounded cooler[4] and he was not wrong.

There are a few limitations we have to be clear on though. First, Cathy will hang on to her properties until she is put into superposition with Alice. That means we cannot turn Bob into Cathy and still have the original Cathy preserved. If we want to transfer Cathy's properties we have to detach them from her. This is called the 'no cloning principle' and says that quantum information can be transferred but never duplicated.

Second, we cannot transfer information without a measurement apparatus at the other end to detect it. Once we have done a measurement on Cathy and Alice, we still have to send a regular signal telling the person with the Bob particle which property to measure. If eye colour is the one that was sent (and we do not have control over this), then it is no good measuring Bob's hair colour, which will have already collapsed.

Obviously the real teleportation process is not measuring things such as eye colour, it is measuring things such as spin state and energy, but these properties are how a particle identifies itself so it might as well be the same thing.

If you do these teleportations enough times you could theoretically transfer every particle property one by one to another particle on a satellite. And if the particle on the satellite becomes identical to the one you started with on Earth, you have effectively teleported it.

DID I MENTION QUANTUM MECHANICS TRAVELS IN TIME?

Kind of. The experiment in question is called the 'delayed choice quantum eraser' and it belongs in the forbidden section of Hogwarts library for corrupting the minds of innocent young physicists.

Create an entangled pair of particles and send one of them (Alice) off to a simple double-slit apparatus with a screen on the other side. Then, separately, send Bob off towards a particle detector that will collapse his wavefunction if switched on, but leave him in a superposition if switched off.

Because Bob is entangled with Alice, what happens to him at the detector will immediately affect what happens to her at the double slit. If the detector is switched on when Bob approaches, he will collapse, forcing Alice to do likewise and she will go through only one of the slits as a classic particle. If the detector is off, however, Bob will carry on as a probability wavefunction and Alice will go through both slits at the same time equally uncollapsed, until she hits the screen somewhere in one of the zebra stripes.

Do this a few hundred times and you will end up with a perfect match. If the Bob detector is switched on 15 per cent of the time, 15 per cent of the Alice particles hit the screen in a classical way, while the other 85 per cent end up 'going zebra'.

If you put the double slit next to where Alice is generated by the entangler (a word which is starting to sound like a pretty awesome *X-Men* villain) Alice will pass through it and be forced into either particle or wave behaviour, depending on whether Bob is detected or not. But what if Bob's detector is a kilometre away?

When Alice gets to the double slit, Bob has not yet reached the detector and discovered if it is switched on or off. If Bob *is* detected, she has to go through the double slit as a particle, but if Bob *is not* detected, she has to go through as a wave. But Bob has not decided so Alice does not know if the detector is switched on. What does she do?

We have delayed Alice's choice of whether to collapse or not until after she has to make it, so the effect is being forced to happen prior to the cause. Surely a machine like this is madness? Well, in 1999 Yoon-Ho Kim built one. And Alice got it right every time.[5] How in the name of Niels Bohr's football-chewing ghost is that possible?

If the experimenters set the detector to measure Bob on 42 per cent of the experiments, 42 per cent of the Alice particles go through as particles. If the Bob detector was on 89 per cent of the time, 89 per cent of the Alice particles go through. No matter how often the detector was on, Alice somehow made the right call.

It is as if Alice sees the future and knows what Bob is going to communicate through their entanglement link. Entanglement information can apparently travel faster than instantaneous. So, can we send messages to the past from the future?

Say we fired three pairs of entangled particles through our experiment and watched the Alice electrons. We create a Bob pathway that is so long that Bob does not reach it for twenty-four hours. One of the scientists agrees that in twenty-four hours she will go to the Bob detector and switch it off and on in a specific pattern to describe the weather. ON-OFF-ON means the weather is sunny whereas OFF-ON-OFF means the weather is wet.

The Alice particle will hit the screen in a corresponding way, telling us which order the experimenter is going to set up the Bob detector in the future. We would have successfully sent a message backwards in time.

Sadly, there is a catch once more. We cannot tell by looking at an individual Alice particle whether it went through both slits or one. Each particle hits the screen in a random location, which could be the result of either classical or quantum behaviour. We only observe the stripy quantum effect when we look at thousands of Alice/Bob pairs and compare the percentages.

This means the time-travel effect is only observed at the end of the experiment and not during. We only see quantum effects if we erase our knowledge of what Alice is doing on each individual attempt. Hence the name: delayed choice quantum eraser.

This phenomenon, whatever it is, is happening inside a metaphorically closed bo . . . chicken and we only see it after the fact. To be honest, quantum mechanics is a flirtatious tease.

THE WORLD SHOULD NOT MAKE SENSE

In the Copenhagen interpretation, quantum effects are said to dominate up to a point called the 'Heisenberg cut', after which classical physics takes over. Anything smaller than the Heisenberg cut will obey Schrödinger's equation and anything larger will obey Newton's. This is physics speak for 'we have no idea what is going on'.

The problem is that, thanks to entanglement, the Heisenberg cut cannot really exist. If you start applying quantum mechanics to a single particle you can easily apply them to two, three, four or however many you want. Entanglements can join any number of particles together so it should be possible to describe Schrödinger wavefunctions for whole people, populations and planets.

We should be seeing quantum craziness all the time, but we obviously do not. This was one of the big five questions we met earlier and is actually one of the ones with which we have made decent progress in recent years.

Consider a measurement on an entangled particle which determines a spin state e.g. up-spin. Since its entangled partner particle instantly becomes down-spin, the two no longer have a bond. Entanglement is broken since they are not governed by the same wavefunction and we can describe them as independent eigenstates.

Now, consider the act of measurement itself. What happens when we measure a particle is that the particle becomes entangled with a particle in our detector, severing connections with any particle to which it was previously entangled.

Measuring one particle of an entangled pair is swapping one entanglement for another, which means measuring something is ultimately the same as entangling with it. Even if we are measuring a single particle.

If a lone particle is in a superposition of up/down spin prior to measurement, effectively it is entangled with itself. Its two possible outcomes are linked via the same wavefunction and when we take a measurement we are breaking that self-entanglement.

SCHRÖDINGER'S CAT IS SAVED/KILLED

If we flip a coin we think of there being two outcomes: heads or tails. But there is technically a third possibility. The coin could land on its side, spinning halfway between heads and tails.

Now imagine slowly bringing your finger towards it as it spins. As soon as you touch it, it will collapse to one side or the other, ending the precarious dance. The coin represents a particle with two possible states, the spinning represents a superposition and our finger stands for a measurement apparatus collapsing the wavefunction.

So far, so Copenhagen. Except the analogy is flawed. The finger that touches the coin is not a different type of object. It is a detector made of particles, bound by the same quantum laws as the coin. So, instead of a finger, we should visualise the detector as a new coin being spun across the table towards the one we are interested in. Both the detector and the particle are in a superposition and when they meet, they slam together and both collapse to an eigenstate.

But if we get the speeds of rotation just right, the coins could theoretically mesh and remain in a combined superposition as spinning dance partners – an entangled pair. For real coins this does not happen very often but for particles it happens easily, provided their wavefunctions are aligned.

Imagine how difficult it would be to get a hundred coins spinning together though, twirling in one great big entanglement. Even if you did get it to happen it would be a fragile arrangement. A single rogue coin in the mix or an external coin entering the system could collapse everything. The more particles you have interacting, therefore, the harder it is to get them in superposition.

There is not a specific number of particles that suddenly makes a system classical. It is just that classical objects have so many particles it is almost impossible to get them all in phase at the same time. The classical world arises because superpositions are unstable but there's no law against saying we could not put a big object into superposition.

A cat might contain trillions of trillions of particles and if you get each to entangle in perfect synchrony, the superpositions could apply to the whole cat. You are unlikely to see that happening, however, because a single air molecule touching the cat would crash the whole thing.

Schrödinger argued that a cat could not be alive and dead simultaneously because it was metaphysically impossible. He was right, but for the wrong reasons. The radioactive particle in the box can be in superposition, but as soon as it entangles with something bigger it becomes less and less likely to hold on to the superposition state.

If we did somehow manage to get every particle inside a cat perfectly coherent and entangled though, the superposition of dead/alive would decohere as soon as it interacted with the box itself. The only way to keep the cat alive and dead at the same time would be to isolate it from its surroundings. Which is what the physicist Aaron O'Connell did in August 2014.

BIG QUANTUM

The 'cat' in O'Connell's experiment was actually a diving-board-shaped piece of metal sixty thousandths of a metre wide, roughly the size of a human hair. The board was suspended above a miniature swimming pool, placed in a box and cooled to a few degrees above absolute zero in order to stop particles inside interacting with each other out of phase. Any random vibration in the material will collapse everything but if all the particles are cold, the object acts as one big particle.

The board was then wired to a circuit outside the box measuring the amount of current flowing through, thus allowing O'Connell to detect how it was behaving without opening the box directly. Once it was in this cold state, O'Connell switched the machine on and sucked all the air out to prevent entanglements forming. Quantum magic ensued.

The diving board began to vibrate gently and vigorously at the same time. The particles were simultaneously moving a lot and a little, meaning every few nanoseconds the atoms were in two places at once, near the equilibrium position and displaced a long way from it. O'Connell had built the world's first quantum machine.

In May 2018 Michael Vanner scaled things up by building a quantum drum that could be both struck and not struck at the same time. In his experiment a 1.7 mm membrane (about the width of a grain of sand) was placed in the pathway of photons, which were given a choice to hit it or not. In superposition, both will happen, meaning the drum will vibrate as it absorbs momentum from the photons and also remain untouched as photons choose the less rhythmic pathway.

The vibrations are too small to be seen by the human eye – only a few photons are actually hitting the drum per second – but Vanner's sensitive apparatus was able to detect photons going along both paths, meaning the drum vibrated and stayed still. Vanner's experiment is even more remarkable because it works at room temperature.

The biggest quantum phenomenon observed, however, was done the year before, by accident. In 2017 a team of researchers led by David Lidzey were experimenting on *Chlorobaculum tepidum* bacteria specimens, shining laser light on them inside a mirrored box in an attempt to affect electrons in their photosynthetic cells.

What they did not realise, and was pointed out the following year by Chiara Marletto,[6] was that photons inside the laser light were entangling with the bacterial electrons, putting the bacteria into a superposition with the light beams. This was possible because the box was so flooded with light that there was nothing else for the bacteria to interact with, meaning their entanglement links to the lasers could survive for a reasonable amount of time. Quantum phenomena can apparently apply to living things.

We are alive at a truly amazing time in physics history and what is about to happen is unprecedented. Quantum mechanics has always been assumed to be confined to the world of the very small, the microscopic, the world of Marvel's Ant-man. But in the last few years, we have started applying quantum rules to objects of the everyday macroscopic world. The age of big quantum has begun.

CHAPTER TEN

Quantum Mechanics Proves I am Batman

More than One Way to Skin Schrödinger's Cat

When trying to grapple with understanding quantum mechanics, the big question is always: can we do better than the Copenhagen shrug of 'it just happens'?

The majority of physics textbooks teach the Copenhagen interpretation because Bohr was very much the big cheese and for decades his way of doing things was the only way. But nowadays, the Copenhagen interpretation is no longer the only game in town.

Obviously we have to abandon classical ideas sooner or later, and any interpretation of quantum mechanics is going to involve some serious weirdness, but since ye saintly days of Copenhagen, a smorgasbord of alternative approaches has been developed.

It is hard to know which interpretations of quantum mechanics to include in a book such as this and in truth they could fill a library. But I decided to follow my gut and discuss three quantum perspectives that were invented by sci-fi enthusiasts, expanding on work of earlier physics giants.

Forget What I Said Earlier

In 1927 Louis de Broglie gave a lecture at the fifth Solvay conference in Brussels. These conferences were physics get-togethers organised by the Belgian industrialist Ernest Solvay who had made millions in the 1860s inventing an industrial method to manufacture sodium carbonate (a key ingredient in glass making).

Solvay envisioned his conferences as summer camp for scientists. Get the world's smartest people together in one building for a month and let them thrash out the big topics. Lectures would be given, debates would be organised and if conclusions were not reached, pointy sticks would be distributed.

The first conference in 1911 was on Planck and Einstein's quantum theory. The second (1913) was on the structure of matter, the third (1921) was about atoms and light, the fourth (1924) was on electricity and the fifth was dedicated to the Copenhagen interpretation and whether it should reign forevermore. In attendance at this legendary event were people such as Schrödinger, Heisenberg, Sommerfeld, de Broglie, Bohr, Born, Planck, Curie, Einstein and numerous others.

There is an iconic photograph of all the scientists in attendance sitting on bleachers like the nerdiest yearbook photo ever taken. Madame Curie is the only woman, Schrödinger is the only one wearing a bow tie and the chemist Paul Debye is the only one sporting a Charlie Chaplin-style moustache, a style which went out of fashion the following decade (for pretty obvious reasons).

At this conference, the softly spoken and affable Louis de Broglie put forward what he felt was a workable alternative to Copenhagen. He decided he had made a mistake introducing wave–particle duality and that electrons and photons were particles only. They did not have wave character, but were instead surrounded by some background substance that *did* have waves in it. The particles were pushed around by these invisible 'guide waves' and thus would appear to move in wave-like trajectories.

According to legend, as he outlined his idea he was heckled loudly by the brash Wolfgang Pauli who had a reputation for interrupting speakers if he did not feel their lecture was up to scratch. Pauli was a superb physicist in his own right (he developed the theory of entanglement we spent the last two chapters explaining) but he was an intimidating presence and de Broglie was basically a nice guy.

De Broglie took Pauli's interruptions with dignity, admitting there were flaws in his hypothesis, but after the lecture was over people were more interested in Pauli's questions than de Broglie's answers and the guide-wave idea was subsequently forgotten.

It was not until 1952 that it was dredged up by David Bohm, a nuclear physicist who had discovered his love of science via sci-fi magazines as a boy[1] and worked on the Manhattan project during the Second World War.

Throughout most of his adulthood, Bohm was a Copenhagen supporter but after being cajoled by Einstein, he began to feel it required too many leaps of faith and turned to de Broglie's guide-wave theory instead.

It even seemed to have some decent experimental evidence going for it. If you send a water wave towards a double slit the same way Thomas Young did, you obviously get an interference pattern, but if you put a tiny object such as a droplet of oil onto the surface of these waves, it will ride them like a boat in a storm, carried towards one of the zebra stripes at the far end. It is still a particle but its final destination is determined by the guiding wave.

The challenge he faced was that this system would mean that the electron should follow the same contours each time, but in a real quantum experiment the electron can show up in any of the zebra stripes seemingly at random. To get around this, Bohm proposed that when each electron is launched from an emitter, it has hidden variables inside it – tiny variations in energy we cannot detect, but which lead to different paths taken at the double slit.

Mathematically, the Bohmian view of quantum mechanics has an extra layer of complexity because you have to talk about the values of these guide waves (called quantum potentials) and that requires more equations on top of Schrödinger's. But in the plus column, it does answer why a particle's properties appear to get decided on detection – properties such as location and spin come from the guide wave instead of the particle. That is why it looks as if particles do not have properties sometimes, they genuinely do not – the guide wave has them.

Bohm, Sweet Bohm

According to the de Broglie–Bohm view of things, quantum behaviour is not really random because you could, in theory, explain the double-slit results with classical physics. In 2010, Yves Couder created an experiment that claimed to show precisely that.

Couder built a double-slit experiment in a water tank and put tiny droplets of oil onto the surface to see how they moved. The oil droplets represented particles and the ripples in the water were the guide wave.

Couder reported that the oil droplets followed the landscape of the water waves and accumulated at the other end in clumps, just like photons and electrons did.[2] Was it possible that particles had locations after all, and they were just riding guide waves like classical objects?

It was an exciting conclusion and seemed to finally knock the Copenhagen interpretation off its throne, but it was too good to be true. Further research by physicists John Bush and, appropriately enough, Niels Bohr's grandson Tomas, failed to recreate the results,[3] and they concluded that Couder had made a few honest mistakes in his setup.

If you try recreating the double-slit experiment with actual particles on actual waves, you get classical results and no zebra stripes. You cannot explain the results of the double-slit experiment using Bohmian mechanics unless the guide wave is some special kind of wave that behaves in a non-classical way. That might be the answer, of course, but all that does is shift the weirdness away from the particle and stick it into a guide wave which, conveniently, we can never observe.

THE HANDSHAKE

The transactional interpretation, which we will look at next, wipes the floor with the Copenhagen interpretation and sticks a finger up at common sense, but it is deliciously cool so it is worth having a look.

This time the original idea came from Richard Feynman, who pointed out that at the quantum level physics works just as well forwards as backwards. A particle moving to the left is the reverse of a particle moving to the right and both processes are equally permitted. Particles do not have a preference for which way time is flowing. An electron emitting a photon can be viewed just as neatly in reverse as an electron *absorbing* a photon. Both are equally real events.

Fifty years later, the physics professor and sci-fi author John Cramer decided to take Feynman's idea and move forwards with it. And, I suppose, backwards as well.

A particle's behaviour is described as a wavefunction but, remember, it has to be multiplied by itself to get an answer. Cramer wondered if the reason we need two identical wavefunctions is because there genuinely are two of them but we only see one because its partner is moving backwards in time.

Let us say our particle is coming up to a double slit. Its normal wavefunction (unfortunately called the 'retarded' wave in the literature) moves forward through time, scoping the possible paths it could take. But at the same time(ish), particles in the detector screen are sending backward wavefunctions (called 'advanced' waves) towards our particle from the future. Whichever future particle happens to send the strongest backward-time signal is the one with which our particle ends up interacting.

Cramer imagines it like a business exchange in which a particle emits an offer and the detector emits a confirmation. The particle approaching the slit and the particle in the detector synchronise wavefunctions in what he calls a quantum handshake, leading to an entanglement between past, present and future.

Delayed choice quantum eraser experiments, those weird ones where Alice knows what Bob is going to do in the future, are suddenly easy to explain. A particle can tell whether it is going to be detected in the future because it gets a message from the future telling it so.

Interestingly, Cramer has said he does not think his interpretation gets rid of human free will.[4] He uses the analogy of a debit card being used to pay for something in a supermarket. The offer wave is the card and the confirmation wave is the bank, but you still get to choose what to buy.

You could, however, make the counterpoint that when you thought you were choosing to buy something such as almond milk, you were actually getting signals from those almonds in the future telling you what to do. Perhaps every decision you think you make is actually the result of a future event guiding your present life choices. Did you really choose to buy this book or was I influencing you from this chapter? WoooOOOOoooOOOooooOOOO (spooky ghost noise).

And I Will Always Love Hugh

The same year David Bohm published his guide-wave theory, Erwin Schrödinger was giving a lecture in Dublin on why he still did not accept Copenhagen. In this talk he explained that although he would sound like a lunatic, his equation was never intended to describe a situation where particles chose a property when measured.[5]

In a superposition, two sets of properties hover around the same particle, that much is known. But where do we get this idea that we have to chuck one of the properties away the moment a measurement is taken? There is nothing in the equation that says that has to happen. In fact, if we take the equation (which always seems to work) literally, it tells us that after a measurement both outcomes are still there, even if we cannot see one of them for some reason.

The Schrödinger equation is a smooth one which describes things evolving gradually, but when a particle interacts with a detector, Bohr insisted we suddenly switch physics and start using chunky particle equations instead. Why should that be?

If we trust the Schrödinger equation it says nothing of the sort. All outcomes happen and there is no such thing as wavefunction 'collapse'. When we measure an electron as up-spin, the down-ness still exists somewhere. We just have to figure out where it is hiding. Enter Hugh Everett III.

A chain-smoking science-fiction obsessive, Everett had a Bachelor's in chemistry and a Master's in mathematics, and decided he should also get a PhD in physics to complete the trio. Under the supervision of John Archibald Wheeler, Everett was charged with trying to come up with a new version of quantum mechanics that did not involve probability.

Everyone in the Copenhagen club was enamoured with randomness, but Wheeler wanted to fight back so he put the challenge to his brightest student and was not disappointed. Everett came up with an answer that not only did away with chance, it also solved the measurement problem.

When a particle is measured in an experiment, all the possible outcomes are actualised. The one we observe gets recorded in our lab book but the other possible outcomes are still there. They just exist in different universes.

Everett figured that superposition was the main headache of quantum mechanics so he eliminated the idea by having overlapping realities instead of overlapping properties. When a particle is offered a choice, the universe splits and each choice is taken in parallel by a different parallel version of the same particle.

A superposition was not really a particle existing in contradictory ways, it was whole universes sitting on top of each other like images on tracing paper.

Provided a particle does not entangle with its environment too much, all the universes involved in a quantum experiment will remain in contact. But as soon as an entanglement takes place (with a detector screen, say) they diverge and become independent realities.

In one universe a particle could go through the left slit and in the other universe it could go through the right. Both outcomes remain mixed together, giving us an interference pattern mid-air, but when the detector screen is reached, every version of the particle hits the screen in a different place – each in its own world.

If you measure the slits with your 'camera' you are not really finding out which slit the particle chooses, you are learning which universe you are in. So when you measure a particle coming through the left slit, a parallel version of you is also measuring it through the right. In Everett's view, known as the 'many worlds interpretation', we no longer have to deal with probability and measurement. We just accept that we are seeing a single slice of a bigger universal cake.

Imagine that 40 per cent of parallel universes have a particle destined for the central stripe of our zebra pattern. Logically, you have a 40 per cent chance of being in one of those universes but you do not know for definite at the start of the experiment where you are. All you can say is there is a 40 per cent chance of the particle hitting the centre of your screen.

The electron is not deciding where to go at random at all. Parallel electrons are going everywhere at the same time but because we only see a single page of the options it looks like an unpredictable outcome. This is a lot more elegant than the Copenhagen interpretation because we do not have to chuck out half the answer for no reason. We simply acknowledge that it is happening in another plane of reality.

THE CAT FINALLY LIVES

This is good news at last! When your cat is in a superposition of dead/alive that really means it is alive and dead in parallel universes. If you happen to open the box and find a kitten stew, you do not need to be sad because an alternate version of you is finding the cat alive and well.

The many worlds interpretation also accounts for the EPR paradox while allowing special relativity to survive untarnished. When two entangled particles are sent to opposite ends of the solar system their properties are pre-decided, just as Einstein believed, but there are two universes with opposite outcomes.

In one universe Alice is up and Bob is down but in the other universe, Bob is up and Alice is down. Prior to measurement we do not know which is which, so we consider the outcome random but that is because our universes have not decohered yet. When we take a measurement and find ourselves to be in the Alice = up universe, both universes split, meaning the universe where Alice is down and Bob is up gets carried away somewhere into the multiverse.

Nobody knows how this splitting works but it seemingly occurs every time a particle is given an option about what to do. In Everett's opinion, measurement was not the issue, choice was.

Right now, as you sit reading this, particles in your body are being given choices about which way around their nucleus to vibrate. In one universe they pick one direction and in another universe they pick the opposite. If you consider how many particles there are in the universe now, and how long it has been around for, the sheer number of choices that have been made is so enormous it is impossible to even give it a name.

Every nanosecond your body, just by existing, is causing trillions of universes to tear away from each other and decohere for ever. The number of parallel realities is so big nobody has even attempted to do a calculation on how many there would be today. Which brings us to . . .

THE MOST IMPORTANT PART OF THE BOOK

Over the years such luminaries as Richard Feynman and Stephen Hawking expressed their support for Everett's many worlds interpretation[6] but, when it was unveiled, Everett was met with ridicule and chose to abandon the whole thing to work for the intelligence division of the Pentagon.

When he died, he insisted his ashes be tossed into the garbage because he was completely unsentimental[7] and, besides, even if he died in this universe he was still alive in numerous others.

We can make that statement because every pathway gets chosen in at least one reality, so different brain processes and events get picked in different worlds and thus every parallel universe has its own version of history. Anything you can imagine will probably have happened somewhere.

The laws of physics would still hold true, of course. There will not be a universe where human skin is made of clouds and frogs glow when you whistle at them. But provided you stick to the standard rules, everything is happening at least somewhere.

There is a universe where America did not win the War of Independence. There is a universe where the Berlin wall is still standing. There is a universe where the Osmonds stuck with rock music after 'Crazy Horses' and never returned to sugary love ballads. And, most crucially of all, there is a universe out there, far into the distant reaches of multi-reality, where I, Tim James, am Batman.

DECISIONS, DECISIONS

The debates rage. At the time of writing there are no experiments that distinguish one interpretation of quantum mechanics from another, so none of them have much right to call themselves definitive.

The Copenhagen interpretation was the big tamale for most of the twentieth century and is still the most popular. But it is also the most infuriating since it requires you to add extra stuff and accept several things on faith.

The de Broglie–Bohm interpretation is the most mathematically complicated and forces us to include hidden variables and hidden guide waves, but if true it would give us a classical explanation for the measurement problem and that is enough to keep it in the loop.

The transactional interpretation is unique among all interpretations because it is the only one to explain why the Schrödinger equation needs two wavefunctions rather than one, but it does not predict a universe where I am Batman so we should reject it on those grounds alone.

The many worlds interpretation does not require you to add anything new to the equations (just new universes) and is by far the most elegant. But is it too much to stomach the idea that all of existence can be splitting constantly into parallel versions?

Picking which interpretation is true is not yet possible. We have equally valid hypotheses so endorsing one over another is a matter of preference and nothing more. But that does not mean we cannot be allowed to have them.

In 2013 Maximilian Schlosshauer asked a group of thirty-three quantum physicists at a convention which interpretation they favoured.[8] Fourteen picked Copenhagen, six went for many worlds (in this universe at least, in another they chose differently), nobody went for transactional or de Broglie–Bohm and a few voted for options we have not covered.

Four people gave no answer and maybe they were the purest scientists of the bunch. After all, if a question is raised for which there is no evidence one way or the other, the honest thing to say is 'I do not know.'

But, as Isaac Asimov once pointed out, humans are emotional beings as well as intellectual ones[9] and, provided we are not committed to saying our choice is absolute, there is no harm in having a favourite. So, pick whichever one makes your brain hurt the most and go with that.

CHAPTER ELEVEN

Far Afield

To Ask a Difficult Question

We have been using the word 'particle' throughout this book with casual abandon so far. But it is time we got specific and nailed things down. If we are going to talk about particle physics seriously we need to decide what exactly we mean by a particle.

The neatest definition physicists sometimes use is 'something which holds itself together and does not fall apart spontaneously'. Your body is a large particle in this sense because your arms do not detach at random (usually) so you fit the definition.

Your body is a composite particle, of course, because while it tends to stay together it *can* be broken into smaller particles – your organs – which also hold together. Organs can be broken down into particles called cells, and then they can be broken into molecules, then atoms and finally protons, neutrons and electrons.

But an electron, as far as we can tell, is not composed of anything smaller. It is not made of sub-electron particles, yet holds together. It is a very special type of particle, distinct from atoms, molecules and cells. It has no substructure and is therefore truly *fundamental*.

Fantastic Mr Faraday

The story of fundamental particles starts during the 1800s with the greatest scientific showman of the age, Michael Faraday. Faraday had grown up the poor son of a blacksmith but believed science should be made accessible to anyone curious enough to pursue it. To that end, Faraday began delivering public talks on science at the Royal Institution, dazzling his audiences with chemical reactions and physical phenomena, and it was at these lectures that he first revealed the discoveries he had made about magnetism, a force like no other.

Magnetism can act through a vacuum and attract or repel other magnets without anything in between to communicate the signal. Magnetism can also work through solid barriers, which is peculiar as all other forces have to be in direct contact with whatever they are pushing or pulling.

Today nothing about magnetism seems remarkable because we live in a world of mobile phones and wi-fi routers, but in the 1800s the notion of causing an effect on something without touching it was tantamount to witchcraft. In order to explain how magnets could do their thing, Faraday urged people to think of magnetism as being made from a 'field' rather than matter itself.

Every magnet creates an invisible distortion in the geometry of the space surrounding it, which other magnetic objects can lock onto. These regions of distorted space dictate the behaviour of particles moving through them, but are not themselves composed of matter.

A field is therefore a non-material, fluid-like substance that tells particles how to move as they pass through. It can be stronger in some places than in others, but you cannot see it directly, only its influence.

The idea is a bit of an eyebrow raiser because we are used to stuff being made from other stuff, and a thing existing without being made of stuff at all is not easy to stomach, but that is the way a field seems to be. Empty space, it would seem, can have properties.

Which reminds me of a joke: did you hear about the scarecrow who won a Nobel Prize? He was out-standing in his field.

I am not going to apologise for that joke, just as Michael Faraday made no apology for introducing fields to physics. There is just no way of accounting for things such as magnetism, electric charge or gravity without fields. And what's more, fields can talk to each other.

ONE FIELD OR TWO, VICAR?

If you take a magnet and waggle it up and down you create a vertical ripple in the magnetic field, but that is not all you create. Faraday discovered that in doing so, you simultaneously generate a disturbance in the electric-charge field, at right angles to the original magnetic disturbance.

This electric field ripple kicks back against the magnetic field, however, which returns the favour by shaking the electric field once again, which agitates the magnetic field in reply, back and forth indefinitely.

The ripple you initially created in the magnetic field is thus carried along by both the electric and magnetic fields vibrating into each other at cross purposes.

We call these ripples 'electromagnetic waves' and represent them in diagrams as ribbons rising and falling, with the electric and magnetic field values pointing at ninety degrees, as shown in the diagram below (the electric field is rippling on the vertical axis E, and the magnetic field is rippling on the horizontal axis M):

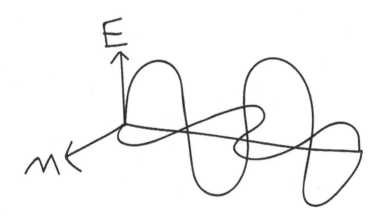

It seems impossible to separate the magnetic and electric fields because when one vibrates so does the other, so sometimes we refer to them collectively as a single 'electromagnetic field' rather than two inter-locking fields.

As you can imagine though, the mathematics needed to describe these overlapping fields is fiendishly complicated. Every point in space needs to be assigned a value telling us how that point will influence an incoming particle.

We imagine tiny arrows (called vectors) at every point of the field pointing in the direction a particle will be pushed. Then, we have to include an electric-field vector at right angles to each magnetic one, which also gets a say in an incoming particle's movement.

Now just imagine what happens if you move one of the magnets and create a wave in the electromagnetic field. Keeping track of an infinite number of arrows all changing direction and size, rotating at right angles to each other, is no simple task, especially for Faraday who never studied maths in school and could not do much beyond simple fractions.

It was the young Scottish physicist James Clerk Maxwell who leapt to Faraday's aid, just as people were starting to call field theory into question, and worked out the maths needed for his fields to be tested properly.

Maxwell's equations helped predict how electric and magnetic fields would interact precisely, and the equations matched the data. Faraday's feelings about fields were finally given fidelity, which was frankly about freaking time.

The Maxwell equations also made a startling prediction. It turns out that electromagnetic waves ought to propagate through space at a very specific speed: 299,792,458 m/s. Look familiar? It is, of course, the speed of light.

Light Me Up

One Saturday evening in 1846 (11 April to be precise), Faraday was accompanying his friend Charles Wheatstone to the Royal Institution lecture hall. Wheatstone was due to deliver a public lecture but decided to have a panic attack instead and ran out of the building, leaving Faraday on his own. In order to avoid disappointing the audience Faraday decided to wing it and improvised a lecture of his own about an idea he had recently been pondering.[1]

He speculated that as particles dance around inside an object they generate electromagnetic waves which travel outward into space until they get intercepted by our eyeballs, notifying us to the particle's presence.

The match between Maxwell's equations and the measured experimental value of lightspeed is too close to be a simple coincidence.

Faraday's guess was apparently correct. The medium responsible for light waves (which both René Descartes and Thomas Young had argued for) was the electromagnetic field itself.

If we vibrate an electron (a particle with electric and magnetic properties) we can create a wave that travels outward at the speed of light. We would not see it, however, because the beam would be too low in energy for our eyes, but a radio antenna might be able to pick it up.

If we vibrate the electron faster, millions of times a second, the electromagnetic distortions would become visible, making the electron glow red then orange, then yellow, green, blue, indigo and finally violet.

If we got it going faster still, the waves would become so energetic they would become invisible once more, bypassing our eyes the way a high-pitched dog whistle bypasses our ears. The electron would now be emitting ultraviolet and X-rays.

In a very real sense, devices such as mobile phones, radio transmitters, wi-fi hubs, Bluetooth emitters, microwave ovens, infrared remotes and X-ray scanners are all just torches. The frequency of their light (a measure of how fast their wavefunction is oscillating) might be too low or high for our eyes to see, but all electromagnetic waves are the same thing.

There is no conceptual difference between a torch beam passing through a sheet of glass and an X-ray beam passing through a sheet of human skin. It is the energy of the wave and the gaps between electron shells of the material, which determine whether the wave is reflected or goes through. The principle is identical in both instances.

Humans see a mere fraction of the colours carried about in the electromagnetic field and Faraday opened our eyes to how blind we are. The stars you see at night are not just emitting visible light; they are also giving out radio, microwave, infrared, ultraviolet, X-ray and gamma waves too.

The very fact we can see the stars at night also tells us that the electromagnetic field must exist in space as well as on Earth, otherwise there would be no medium to transfer the energy to us. The electromagnetic field floods the cosmos from edge to edge. You're sitting in it right now.

When your eyes perceive words on this page it is because electrons in the page's surface are jumping between energy levels, disturbing the electromagnetic field around them. These electromagnetic disturbances move through the field towards your face and finally reach your eyeballs where they are absorbed by electrons at the back of your retina, causing them to send currents down an optic nerve to your brain. Vibrations in the electromagnetic field are, in a very literal sense, the only things you have ever seen.

Ice Cream and Bed Sheets

This is all very classical so far, what with fields rippling smoothly, etc. But we know, thanks to Planck and Einstein, that light energy is split into chunks: photons. So, in the late 1920s the reclusive English physicist Paul Dirac decided to invent a way of explaining the electromagnetic field in quantum terms – a 'quantum field theory'.

It is always unwise to retro-diagnose a person as having a certain medical disposition after they are dead but it is very likely Paul Dirac was on the autistic spectrum. He had no concern for social nonsense, no desire for publicity or glamour, understood people's words as they were literally meant and was so reserved his students jokingly invented the 'Dirac unit' to refer to a speaking rate of one word per hour.[2]

So, here is another joke. Did you ever hear about the magic tractor? It went down the road and turned into a field.

That joke functions (barely) because the idea of a tractor literally turning into a field is absurd. A lumpy object such as a tractor cannot be transformed into a smooth field because objects and fields are different things. They cannot be interconverted. But in quantum field theory this is exactly what we do.

A photon is a particle (it holds itself together) but since it has wavelike character (and a wave is an oscillation in a background medium) there must be some field out of which the photon is being vibrated into existence.

In classical terms we think of electromagnetic fields undulating up and down, but in quantum terms we think of energy as an isolated blip in the field, like a spike on a heart monitor.

The electromagnetic field is a placid backdrop to our universe, but if one particular region gets agitated it can bunch up and form a twisty little knot of concentrated energy. This packet of field disturbance is a 'quantum of the electromagnetic field'. A lump in the backdrop. A photon.

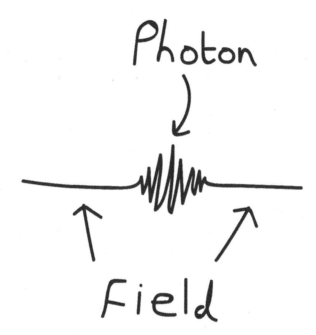

Drawing field quanta in three dimensions is hard so most analogies imagine fields as a flat surface with particles appearing out of them as in the diagram above.

One helpful approach is to picture a smooth bed sheet pulled across a mattress. If we pluck at a point on the sheet we can create a tiny fabric mountain, representing a quantum of the field. A bed sheet particle, if you will.

These field quanta (particles) can be transferred from one point to the next, giving the impression of moving through empty space. We just have to remember that the field is actually a three-dimensional fabric that surrounds us in all directions and photons are excited out of it.

I sometimes like to think of it as being analogous to scooping a ball of ice cream from a tub. The flat surface of the ice cream prior to our scoop is the undisturbed electromagnetic field and when we carve little spheres out of it they are the photons we detect in experiments.

Really, we should start calling the electromagnetic field the photon field, but you know how much physicists love clinging to outdated terminology (looking in your direction, 'spin').

You Are Not Made of Particles

Paul Dirac was able to explain mathematically how we could extract photons from the photon/electromagnetic field, and showed that since every type of particle has wave character, every type of particle can be thought of as a quantum in its own field.[3]

There is an electron field everywhere in the universe, overlapping with the electromagnetic field, and when we disturb it we get an electron bubbling out. All particles, including those in your body, are really excited vibrations of invisible underlying fields.

The particles you are made of and the empty space surrounding you are not separate at all. You are made from packets of energy floating through fields made of nothingness. Whether you consider this fact disturbing or beautiful is up to you.

CHAPTER TWELVE

Lines and Wiggles

ONE THEORY TO RULE THEM ALL

Paul Dirac's hope was that quantum field theories would one day be able to explain every conceivable phenomenon in physics. Every particle would be treated as a quantum in a field, and the interactions between fields would underpin the interactions between particles. The idea was a potential game-changer for physics. But unfortunately it was so complicated it turned out to be a game very few people could play.

The mathematics of quantum field theory is difficult. Crazy difficult. There is even a $1 million prize offered by the Clay Mathematics Institute to anyone who can solve one of quantum field theory's tougher challenges (called the Yang Mills Existence Mass Gap Problem if you fancy a go this weekend).

To try and make headway with something so intractable, Dirac suggested we start small and only consider the two simplest particles/fields: electrons and photons. He called the interaction between them 'quantum electro-dynamics', or QED for short, and his hope was that if we could work out a fully detailed QED theory, we could start adding in other particles and go from there.

At the end of his 1930 book *The Principles of Quantum Mechanics*, Dirac concluded with the words, 'it seems that some essentially new physical ideas are here needed'. Quite an understatement, but hardly surprising given the reserved nature of Dirac himself.

His words drifted into the physics community like a gentle challenge on the breeze, and eventually snagged in the mind of a man who was by all accounts his polar opposite. Science's most charismatic and colourful rogue: Richard Phillips Feynman.

LITTLE DRUMMER MAN

Born in New York to a uniform salesman and his wife, Richard Feynman showed enormous talent for physics from his first exposure. His official IQ was clocked at 123 (reasonable, while not astounding) but by the time he was an adult Feynman was regarded as the most gifted scientist on the planet, comparable even to Einstein.

To give you some idea of how clever he was, in 1958 when NASA launched the Explorer II satellite, something failed during the ascent and it never reached orbit. Feynman bet the NASA engineers he could calculate where the satellite would land, faster than their computer. Not only did he win the bet, he calculated the answer more accurately. Twice.[1]

Feynman was also the soul of any party he was invited to and entertained his numerous friends with safe-cracking tricks, bongo-drumming skills and circus juggling. He had red carpets laid out for him at weekly lectures and spent his free time hanging out in topless bars, doing calculations on napkins or drawing sketches of the dancers and sometimes the men watching.[2]

A skilled raconteur and an unashamed prankster, Feynman was the Han Solo of physics. But, above all else, he possessed the finest mind of his generation.

He began his studies at MIT before moving to Princeton (earning a perfect entrance-exam score) to complete his doctorate under John Archibald Wheeler, who also guided Hugh Everett on the many worlds interpretation.

Halfway through his PhD, he was recruited by Robert Oppenheimer to help the American military design an atom bomb and was described by Oppenheimer as 'the most brilliant young physicist here',[3] which is saying something considering 'here' referred to Los Alamos National Laboratory – an establishment set up exclusively to house the smartest scientists in the world.

After the war, Feynman completed a post-doctorate at Cornell and then took a professorship at Caltech where he sought to shed the unpleasant taste in his mouth left from helping with the bomb. He decided to commit his life to three things only: thinking, teaching and taking care of students.[4]

The teaching and taking care of students bit was easy. Feynman's nickname was 'the great explainer' since his lectures were so good they were attended not just by freshmen but senior colleagues who found themselves learning their own subject better from hearing his take on them. That just left the third focus: thinking. And what he decided to think about was Dirac's challenge.

I'M DRAWN THAT WAY

Coming up with a fully detailed quantum field theory for electrons and photons had big problems. Quite literally. A lot of the calculations yielded infinite answers or needed an infinite number of inputs to get to an answer, which obviously cannot be right for a finite universe (see Appendix IV for a slightly more detailed look).

It is hard to say what made Feynman different to other geniuses of the same era, but I personally think it comes down to him being a physicist first and a mathematician second.

Not to undersell him, he was a mathematical virtuoso second to none, but to him the equations were a language and not the final goal. You had to keep a focus on the stuff they described and not get bogged down in the symbols. Since the mathematical language being used by everyone to tackle QED was cumbersome and only generating partially correct answers, Feynman decided to invent a new type of maths to make life simpler.

Start by picturing an electron minding its own business as it travels through the universe. In quantum field theory terms, we have to describe this as a quantum of energy in the electron field propagating from one place to another. We describe its trajectory with an equation called (reasonably enough) a propagator.

In Feynman's new mathematical system, we replace the electron propagator equation with nothing more complicated than an arrowed line.

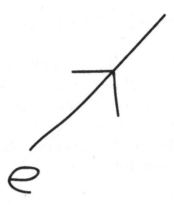

(NB: Strictly speaking this symbol represents 'electron in motion' and does not have to represent a straight line from bottom to top; it could also be going in a curve or circling a nucleus.)

But now, as the electron moves on its intended course, an incoming photon approaches and gets absorbed into it, knocking it somewhere new. In quantum field theory terms we need to describe a quantum of the photon field meeting our electron and an energy transfer taking place between the two fields.

We represent a photon propagator with a wiggly line and draw the interaction like so:

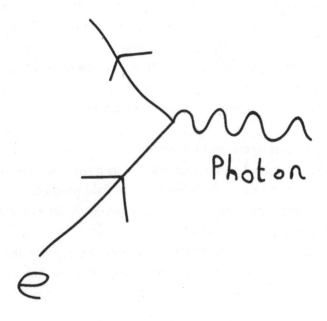

Reading from the bottom we see an electron propagating through its field, interacting with the photon field (absorbing a photon) and then blasting off in some new direction as it absorbs the energy. Or, just as easily, it could be describing the reverse process – an electron emitting a photon and recoiling in another direction, like someone's hand flying backward as they fire a gun.

The point where the three lines join is the 'vertex' of the diagram and is handled mathematically with something called a coupling constant. A coupling constant is just a number that measures how easy it is for two fields to exchange energy. The higher the number, the more likely two quanta (particles) are to interact.

The whole thing looks blindingly simple, but that is where the power of Feynman's approach lies. Feynman diagrams cut out pages and pages of excessive maths jargon and whittle it all down to the essentials. Take the incoming electron propagator, the photon propagator, the outgoing electron propagator and the coupling constant, multiply them together and you get a prediction for how an electron and photon will interact. A quantum electrodynamics theory which worked.

CHARGE EXPLAINED . . . AT LAST

An ordinary beam of light is made of photons obeying certain laws about direction, velocity and energy. But if two electrons pass each other they can exchange a photon like footballers passing a ball and, due to Heisenberg uncertainty, we cannot tell which way the photon actually moves. We can say a photon exchange takes place, but not which electron receives and which donates: that would give us too much information about momentum and position.

These photon exchanges that occur between electrons are like temporary photon ripples rather than permanent beams of light passing between, so they are clearly not your average photons.

Imagine two boats moving across the surface of a lake and passing within close proximity. As they do so, the wakes generated from each boat will meet in the space between, creating a temporary water disturbance that pushes both of them apart. The boats never touch, but this momentary fluctuation between them in the water field allows them to exchange energy, deflecting at angles rather than passing in a straight line.

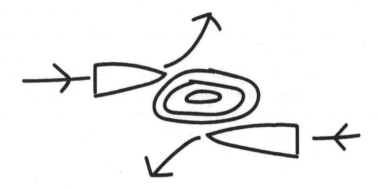

The toy boats (viewed from above in this diagram) represent electrons and the bulge in the water (the concentric circles between them) is the exchanged photon, which exists for only a moment as energy is transferred.

We call these momentary ripples in the electromagnetic field 'virtual photons' to distinguish them from the actual, permanent photons that comprise beams of light. The same way you might call the wave that pushes two boats apart a swelling of the water, rather than a permanent wave, which roams the ocean independently.

Virtual photons do not exist for long and do not have to follow the normal rules of physics, so we can assign them all sorts of properties we would not normally see, to account for any phenomenon we wish.

A virtual photon can transfer energy between electrons forcing them apart, but if one of the particles is oppositely charged, we can give the virtual photon a 'negative energy', which sucks the particles together like a whirlpool.[5]

The diagram for two like-charged particles repelling is shown opposite on the left, and attraction for oppositely charged particles is shown on the right. The calculations are a little trickier for these diagrams because we need to include two coupling constants (vertexes in the diagrams) and five propagators (for each particle line), but the answers generated by QED are bang on.

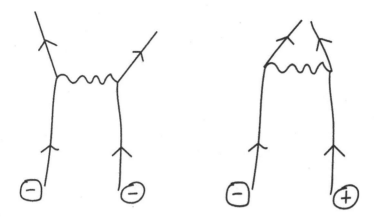

Electric charge turns out to be a measure of how strongly a particle interacts with the photon field and which way the virtual photons are behaving.

The physicist Julian Schwinger (who shared the Nobel with Feynman) said we should therefore imagine an electron emitting and absorbing virtual photons with itself constantly, like a person juggling as they run, thus creating a virtual photon cloud around itself as it moves, and other particles can bump into this cloud. QED explains the very nature of electric charge.

Breaking the Law

One of the things we have been counting on since the beginning of the book is that cause always leads to effect and effect does not happen without cause. You never get something out of nothing nor can you extinguish something into nothing. This principle, laid out by Émilie du Châtelet in 1759, is often stated as 'energy cannot be created or destroyed', the first law of thermodynamics. And it is a law virtual photons are allowed to break.

Heisenberg uncertainty tells us that we cannot take precise measurements on a particle. Momentum and location are never fixed simultaneously so a particle cannot sit still. Quantum field theory extends this and says that whatever is true of particles must be true of their underlying fields, which means a field cannot have fixed values either.

Quantum fields have to jitter constantly and since these jitters are virtual particles, that means every empty field is constantly creating and destroying countless virtual particles every second. Every point of space around you is foaming and frothing with virtual particles sparkling into and out of existence in less than the blink of an eye. Empty space is not really empty.

Feynman calculated that if you take the energy of all the virtual particles appearing inside the volume of a single light bulb, you would have enough energy within them to boil the entire Earth's oceans. We just do not notice all that energy because it disappears almost as quickly as it appears.

This realisation means that in quantum field theories you *can* get something for nothing because 'nothing' is unstable and the uncertainty principle will not allow it to stay put. If you have emptiness for a long enough time, energy will appear without a cause. It might sound like a crazy suggestion but we have to swallow it because QED has something weighty in its favour.

It is the Most Accurate Theory in All of Science

To give credit where it is due, Feynman was not the only person to devise a fully working quantum field theory for electrons and photons. He shared his Nobel Prize with Shinichiro Tomonaga and the aforementioned Julian Schwinger, who had their own methods for calculating QED predictions.

Schwinger and Tomonaga's versions were much beefier though and included a lot of excess work which Feynman showed was not necessary. (Feynman and Schwinger's approaches were so radically different, nobody even realised they were working on the same problem until their mutual friend Freeman Dyson made the connection one afternoon while sitting at the back of a hot bus on his way to Ithaca.)[6]

Feynman diagrams are elegant, but you do not get a Nobel Prize just for drawing pretty pictures. Believe me, I have submitted dozens of my book illustrations to the Nobel committee and have heard nothing. QED did a little better though, because Feynman diagrams are not just fanciful sketches. They have serious predictive power.

One example of QED's strength is the value it calculates for how strongly the photon and electron fields couple to each other (exchange quanta). The most exhaustive calculation of this number was performed in 2012 by Makiko Nio and his team, who computed the outcome of 12,672 Feynman diagrams each containing ten vertexes between photon and electron fields.

Their value of the coupling constant for the fields was 0.00729735256. The value measured in experiments is 0.00729735257. That is an agreement between theory and data to ten decimal places.[7]

Feynman described this kind of accuracy as like measuring the distance from New York to Los Angeles and getting it right to within the width of a single human hair. No other prediction in science even comes close.

If you accept any scientific theory, anything at all from what causes warm air to rise to how viruses work, you should probably accept QED too because its evidence is stronger. And if the numbers do not convince you, there is another prediction QED makes which matters. Or rather, antimatters.

SOUNDS LIKE SOMETHING FROM A MOVIE

When Dirac talked about particles appearing out of fields he pointed out that doing so should leave behind a hole in the field. Going back to the ice-cream analogy, each scoop we make from our ice-cream tub creates an ice-cream particle but also an equal-sized crater in the surface.

We can cancel it out by putting the particle back into the void, but it looks like generating a particle generates an inverted particle-hole simultaneously. An anti-particle.

Feynman's QED predicts anti-particles as well, but they arise in a different way. Rather than one electron field out of which we make an electron and a hole, there are two fields: one for electrons and one for anti-electrons, with the photon field coupling to both.

Let us revisit our Feynman diagram for an electron absorbing (or spitting out) a photon.

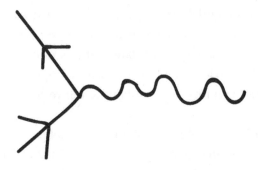

One of the nifty features of Feynman diagrams is that they are valid from any angle, meaning we can rotate them and get an equally correct answer. If we flip the above diagram by ninety degrees we end up with this . . .

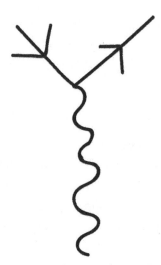

When read from the bottom upward, a quantum in the photon field is propagating through space, then randomly decides to die, transferring its energy into the electron field (a photon turns into an electron). But if we look closer we can see something odd. One of the electrons has its propagator arrow pointing backward.

The propagator on the right is representing the electron, but the particle on the left must be some sort of reverse electron generated at the same time. And we can flip the whole thing again . . .

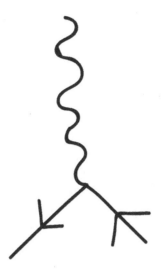

The two arrowed lines now show an electron (on the right) and an anti-electron (on the left) approaching each other and annihilating to produce a photon. (NB: for mathemagical reasons, the collision actually produces two photons rather than one but it does not make any difference to our picture.[8])

'What would an anti-electron be like?' I hear you ask. Well, something like a normal electron but with the charge reversed. A positive electron rather than a negative. But, if an electron gets its charge from juggling photons in one direction, is the opposite charge the result of juggling photons the other way? Feynman, rather cryptically, said yes.

In 1949 he showed that if you take the propagator for a normal electron and flip the direction of time in your equation (reverse the diagram arrow) you get the propagator for an anti-electron. Anti-particles are, according to Feynman, regular particles moving backwards in time.[9]

This idea of a backwards-time electron is scalded by some modern physicists because you cannot seriously claim an electron can time travel. To me, this feels a little like Einstein objecting to superposition. He did not like what it implied but it is not possible to determine if it is right or wrong yet; all we can say is that the equations work. What this means is a matter of taste. Antimatter particles do exist and they behave exactly the way Feynman said they would.

BUILD YOUR OWN PARTICLE DETECTOR

Antimatter was discovered by Carl Anderson using a device called a cloud chamber. The design is so straightforward anyone can build one. I have made several and I am as useless in a lab as Heisenberg, although I do know how to change the battery in a smoke alarm (funnily enough, a smoke alarm is crucial for any lab I am working in).

Here is how you do it. Get a transparent tank and line the edge all the way round with a strip of felt soaked in alcohol (propan-2-ol/rubbing alcohol works best). Seal it back up and stand the whole thing on a layer of ice to cool the bottom surface. This will establish a thin atmosphere of alcohol vapour inside the tank and any particles which zip through the plastic walls will leave a vapour trail in their wake, showing up as pale whispery lines, seemingly out of nowhere.

You can also put a magnet inside and charged particles will move in a curved path around it, charge and magnetism being characteristics of the same field.

Carl Anderson was studying cosmic rays – particle debris that rains down on us constantly from space – and counting the electrons that reach the surface of the planet. When he did so, most behaved exactly as expected, but fifteen of the tracks moved the wrong way around the magnet. Anderson was observing positively charged electrons. Antimatter from space.[10]

Anti-electrons were named 'positrons'. A reverse-charged proton was disappointingly called an anti-proton, however, and a reverse-neutron is similarly called an anti-neutron, although you might be wondering how it is possible to have a reverse-charge neutron – famously a neutral particle. We will get to that in the next chapter.

Thanks to QED, our picture of reality became a lot more complicated because there are now seven particles/fields to deal with: protons, anti-protons, neutrons, anti-neutrons, electrons, positrons and photons.

Photons do not have antimatter counterparts, which actually makes perfect sense in the Feynman time-reversal view. If antimatter really is regular matter going backwards in time, photons should be their own anti-particle because they do not experience time.

As we have already seen in special relativity, time slows down until you hit the universal speed limit and, since photons are already moving at that limit, their notion of time is non-existent. Photons do not age forwards which means they do not age backwards either.

❀

THE WEAPON OF CHOICE FOR COMIC-BOOK VILLAINS

Antimatter particles have a short life expectancy because as soon as they meet regular matter (most of the universe) they cancel out producing photons. But worry not, you can make your very own antimatter particles right here on Earth for the low, low price of $62 trillion per gram![11]

Obviously, because it is so difficult and expensive to produce, antimatter is only ever generated in tiny amounts by particle physicists with a lot of patience. The record for keeping the most antimatter alive at the time of writing is a whopping sixteen and a half minutes for 309 atoms of anti-hydrogen (an anti-proton with an orbiting positron), which was achieved in 2011.[12]

The main reason it might be worth looking into antimatter as a technology is that the energy you get from matter–antimatter collisions is so great you could power a rocket to Alpha Centauri with little more than a teaspoon of the stuff. It could also accelerate a medium-sized spaceship to about a quarter the speed of light, allowing you to make the trip in a matter of years rather than centuries.

That kind of energy does make it ripe for weapons manufacture, of course, and the notion of antimatter bombs has been occasionally discussed by military officials. Depending on how we play it, antimatter could become the most destructive thing on our planet or the very means by which we escape it.

CHAPTER THIRTEEN

Particle Physics Gets Jacked

PARTY-CRASHER

The year is 1936. The atomic structure has been solved and quantum field theory is making accurate predictions. The last time we felt this confident was right before Max Planck began experimenting with light bulbs and surely that could not happen again. How could there be anything else?

Well, as the oft-quoted Robbie Burns poem says: 'the best laid plans of mice and men, do not take into account the presence of the muon field'.

Carl Anderson had discovered antimatter by observing positron trails in a cloud chamber. That was awesome but it had not blown anyone away since antimatter was an expected prediction of QED. What did blow everyone away in 1936 was another trail, spotted in his cloud chamber, which behaved just like an electron . . . only two hundred times heavier.

This particle, christened the muon, has the same characteristics as an electron, couples to the photon field and follows Feynman's diagram rules; it is just fat and, as far as we could tell, completely unnecessary for our theories to work.

No atom contains muons because they are short-lived particles, lasting for about two-millionths of a second, so they are present in the universe but have no apparent purpose, prompting the Nobel Laureate Isidor Rabi, to bellow 'Who ordered that?' in astonishment when he was told there was a new field none of our theories predicted or asked for.[1]

Because they are so heavy they are also high in energy and, in the same way an energetic guitar string dampens down to a softer vibration, fluctuations in the muon field rapidly transfer their energy to the electron field, decaying the heavier particle into the lighter one (a fancy way of saying muons turn into electrons).

Then it happened again in 1974 when Martin Perl discovered the tauon (often just called the tau), an even heavier electron, this time three and a half thousand times heavier with an even shorter lifetime.[2]

Electrons and positrons were not unique it turned out. They were the lightest of members of a particle family comprising electrons, muons, tauons and their antimatter twins. These six particles are collectively called 'leptons' from the Greek *leptos*, which means small, and their existence is a little unsettling.

There was once a time when we imagined every law of physics conspired in some way to permit or encourage life. The discovery of the muon and tauon challenges that view because it would appear nature sometimes does stuff that has nothing to do with us whatsoever. Life exists just fine without muons and tauons. Whatever they exist for, we apparently do not need them.

Muons and tauons have a few esoteric uses such as probing the interiors of pyramids (they are heavier and penetrate deeper than an electron beam would) but other than that, it would appear nature triplicated the electron for no reason. And it does not stop with leptons.

THE PARTICLE ZOO

Cosmic ray particles are hard to detect because most of them interact with our atmosphere and never reach the surface. To get a better view, Cecil Powell decided to stick a bunch of particle detectors at the top of the Andes Mountains and see what was coming down. At these great heights, in 1947, he discovered a particle he called the pion, which had the same charge as a neutron, but a lower mass.

A few months later, the kaon particle was discovered by Clifford Butler in a similar way. Then in 1950 we discovered the lambda particle, which acted like a heavy proton. Then we found xi particles, eta particles, omega particles and, by the early 1960s, there were over four hundred new particles to keep track of.[3]

Our neat collection was looking more like an unruly party, with new gate-crashers arriving every five minutes. Robert Oppenheimer remarked that they should just give a Nobel Prize to any physicist who managed to *not* discover a new particle[4] and Enrico Fermi expressed his frustration by saying, 'if I could remember all those names I would have become a botanist!'[5]

Quantum field theory was supposed to be an elegant, albeit mathematically complicated, description of the fundamental laws of physics. This ugly potion of particles was not painting such a portrait.

It was reminiscent of what happened in chemistry a century before. New chemical elements were being discovered with a variety of properties and the confusion was only brought to order when we realised atoms were made of smaller things – the protons, neutrons and electrons with which we are today familiar. Physicists began to hope that something similar would happen for particles.

Four hundred different species in the zoo looked too messy. Someone was going to have to find a pattern in the chaos, the way Feynman had brought order to our understanding of electrons and photons. Fittingly, or perhaps ironically, the person who achieved this monumental task was Feynman's rival: Murray Gell-Mann.

Gell-Mann's office was across the corridor from Feynman's and there was often tension between them, perhaps worsened by the fact they both had Nobel Prizes.

Feynman was a partying entertainer who enjoyed the company of women (he married three times) and rarely bothered to read books. Gell-Mann was a distinguished academic who attended Yale at fifteen, spoke many languages and spent his time reading papers on linguistics and archaeology. Gell-Mann was about the quiet life whereas Feynman was about the bars and clubs (although it is worth noting that Feynman never drank alcohol and encouraged sobriety).

Despite their disagreements and different lifestyles, both men suspected that protons and neutrons were not fundamental after all. There were scores of known lighter particles, which implied a sub-structure of smaller things, and the race was on to come up with a new quantum field theory to describe them.

Feynman referred to these hypothetical sub-proton/neutrons as 'partons' and did a lot of work on how we might observe them. The theory that described them in detail, however, was laid out by Gell-Mann, who called them 'kworks' for no other reason than he liked the way the word sounded (if you've read about this topic before and are not sure about my spelling, hold on).

By analysing the masses, charges, spins and lifetimes of the abundant particles being discovered, Gell-Mann showed that all of them could be explained as combinations of kworks, which came in two varieties he named up and down.

Up kworks have a positive charge of $+\frac{2}{3}$ while down kworks have a negative charge of $-\frac{1}{3}$. If you combine two up kworks with a down you get $+\frac{2}{3}$, $+\frac{2}{3}$ and $-\frac{1}{3}$ which adds to $+1$ . . . a proton. If instead you combine two down kworks with one up kwork you have $-\frac{1}{3}$, $-\frac{1}{3}$ and $+\frac{2}{3}$, which cancel to zero, giving us the neutron.

Three up kworks becomes a particle called a delta. An up kwork with an antimatter down kwork becomes the pion and so on. The particle zoo was an illusion, and protons and neutrons were composite particles, not fundamental. Kworks were what mattered because they made up matter.

Oh, and that is how you can have anti-neutrons, by the way. While a regular neutron is chargeless, its constituent kworks are not. You can have two anti-downs with an anti-up totalling to zero, but an antimatter zero rather than a matter zero. Ain't life grand?

CRY OF THE SEAGULL

One evening, while reading a copy of the novel *Finnegans Wake* by the Irish modernist James Joyce, Gell-Mann came across a poem opening with the phrase: 'Three quarks for Muster Mark.'

He was immediately struck by this nonsense-word 'quark' because it described something occurring in a group of three, just like his proposed particles often did. The spelling matched the sound he had picked – sort of – and he adopted it from then on. Joyce probably intended the word quark to rhyme with Mark but Gell-Mann decided it should rhyme with quartz instead.[6]

In the poem, the word represents the noise a seagull makes and presumably seagulls in California, where Gell-Mann lived, make a 'kwork' sound rather than a 'kwark'.

In Britain, people usually pronounce the word as 'kwark' but that is not what Gell-Mann wanted. You have to pronounce it kwork or face the wrath of Californian seagulls.

Besides, naming a particle after a bird noise is hardly the weirdest thing out there. The physicist Alan Guth named a hypothetical particle of his invention an inflaton after its supposed ability to inflate the universe, while Frank Wilczek named a particle of his invention the axion after a brand of laundry detergent.[7]

COLOURFUL LANGUAGE

Quarks were discovered experimentally, a few years after Gell-Mann proposed them, in a process that fires leptons (electrons, muons and tauons) at a neutron and watches their paths. If the neutron was a single chunk of matter in a 'neutron field' then electrons would recoil at a sharp angle, but if a neutron was made from a quark sub-structure, as Gell-Mann predicted, the leptons would instead be deflected, pulled off their course by the quarks' partial charges.[8]

The outcome of these experiments matched Gell-Mann's predictions, giving us a new type of particle and a quantum field theory for the nucleus in the process.

Protons and neutrons are composed of three quarks, plus thousands of virtual quarks popping up around them thanks to Heisenberg uncertainty. The three quarks that remain constant are called 'valence quarks' and they determine the overall identity.

We know quarks interact with the photon field since they have an electric charge but the obvious question is why two up quarks, which are both positive, do not repel. Two like-charged particles never hang out together so someone needed to explain why the nucleus of every atom does not blast itself to pieces within moments of forming.

The Japanese physicist Hideki Yukawa proposed a force much stronger than electromagnetism that kept protons in one piece, as well as holding protons and neutrons together. This force could overpower charge repulsion because it was quite strong, so he named it (brace yourself for this) the strong force.

The difference in magnitude between the electromagnetic and strong forces is pretty vast. Electromagnetic interactions are the kind that move electrons around atoms and get chemical reactions going, e.g. they start fires. The energies involved in the strong force are about moving protons and neutrons in the core of an atomic nucleus. They start nuclear explosions.

Since the electromagnetic force is all about particles coupling to the photon field and communicating via virtual photons, logically the strong force should have its own field for the quarks to couple to, which Gell-Mann called the gluon field. Y'know, cos it glues things.

Next, we need a property to go with it. A particle's ability to couple to the photon field is called its electric charge. Gell-Mann needed a name for the property that allowed quarks to couple to the gluon field and he chose the rather unhelpful word, colour.

Stickiness might feel a more intuitive name, but Gell-Mann did have a reason. Unlike electric charge, which comes in two varieties (positive and negative), colour comes in three, reminiscent of the three primary colours of light.

Quarks with red, green and blue 'colours' bind together via gluons and the colours cancel out to make a proton or neutron 'white' overall. Quarks are not literally red, blue or green (they do not have colour actually – see Appendix V) but we usually draw them that way just to confuse things as much as we possibly can.

Feynman's quantum field theory for electrons and photons was quantum electro-dynamics so Gell-Mann called his theory for quarks and gluons quantum chromo-dynamics (QCD for short), from the Greek word *chroma* meaning colour.

Looks Like We're Stuck this Way

The big difference between Feynman's QED theory for leptons and Gell-Mann's QCD theory for quarks is that in QED everything can be described in terms of opposites. Attraction and repulsion, positive and negative, matter and antimatter, etc. All these things can be handled by reversing stuff in your equations and diagrams. But with the strong force there are three kinds of colour so it is no longer a matter of back-pedalling. When there are three varieties available the word 'opposite' does not even apply.

Also, antimatter quarks have anti-colours named anti-red, anti-blue and anti-green, so really there are *six* colour charges we have to explain if QCD is going to make sense. Electric charges are the result of photons behaving in opposing ways but gluons are different and it was their quirky (or quarky) behaviour that Gell-Mann would have to rationalise.

A more complicated type of Feynman diagram was needed, in which colour could be transferred.

Electrons and positrons hold their electric charge but quarks can swap colour back and forth between them. If you have two quarks, let's say red and blue, the gluon field can exchange their colours so red becomes blue and blue becomes red.

The diagrams for QCD use coiled lines to represent gluons and the interaction between two quarks can be calculated/drawn like this:

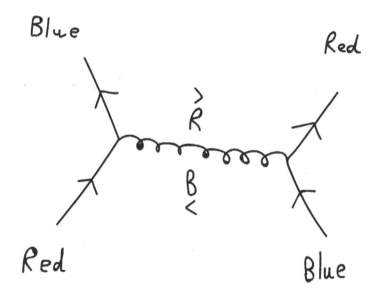

The gluon that moves between them is carrying blue charge to the left and red charge to the right, meaning virtual gluons are multicoloured as they shuttle back and forth.

This colour exchange also explains why the strong force is always attractive whereas electromagnetism can both attract and repel. Since virtual gluons are multicoloured, there always has to be a quark at each end. Gluons, by definition, are transferring colour from one quark to another so if you delete one of the quarks the gluon is left with some colour charge and nowhere to put it.

Quarks have colour, which is another way of saying they couple to the gluon fields and thus they never exist on their own because part of their identity is 'being bound to other quarks via gluons'. The strong force is always attractive.

The term for this is 'quark confinement'. Quarks are always found in pairs (called mesons), trios (baryons), quartets (tetraquarks) and so on. We never observe quarks 'naked' – genuine term – although physicists being perverts have desperately tried to catch a glimpse.

We can take two quarks at the end of a gluon thread (a meson) and spin them inside a magnetic field until the gluon tube snaps.

Unfortunately, when we run this experiment the gluons wind up with excess energy and transfer it instantly into the quark field, creating new quarks at each end of the rip. One meson becomes two. Quarks do not exist on their own even if we try to pull them apart.

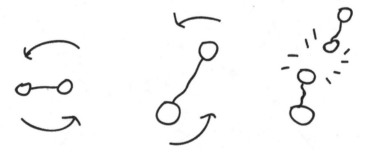

In Feynman's QED, photons do not possess a charge (they cause it) but in Gell-Mann's QCD, gluons *do* possess colour as well as their quarks, meaning gluons will self-interact, swapping quantum rainbows back and forth at will.

Originally it made sense to think of a proton as a trio of quarks with gluons moving between them to form a triangle, but because gluons talk to each other and cling together, the gluon tubes between quarks are now thought to form a Y-shape.

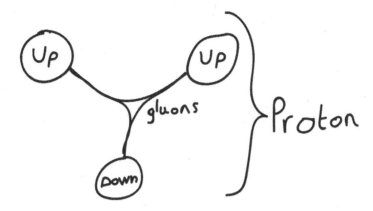

This also means gluons can stick together without needing quarks at all, forming their own tangled gluon–gluon bird's-nest structures called glueball particles.

In some ways, all these additional complications make QCD a more impressive and intricate theory than Feynman's. On the other hand, QED is far more elegant, requiring only one exchange particle while QCD requires several (eight types of gluon to exchange all the colour combinations). Thank God there are only two types of quark. Right?

STRANGELY CHARMING

Gell-Mann's up/down quarks are great. By combining them in the right order we can account for almost all the known particles in the particle zoo. The key word there being 'almost'.

One particle in particular, the kaon, cannot be described as a combination of up and down quarks. It acts more like an up quark glued to a heavier version of a down quark.

Given that electrons were known to have more massive counterparts (the muon and tauon), Gell-Mann decided the down quark must be the same. The behaviour of kaons was undeniably strange, so Gell-Mann used that as his name for the third quark, giving us a list of all required quarks in QCD:

UP	
DOWN	STRANGE

But come on. Look at it.

Gell-Man used complicated mathematics to predict the up and down quarks, but it does not take a Nobel Prize winner to see something is missing there. If the down quark has a heavier partner does it not seem likely the up quark should have one too? Would it not be a whole lot neater and prettier if there was a fourth?

The physicist Sheldon Glashow believed in this fourth type of quark and worked out what properties it was likely to have. There was some minor evidence to suggest its existence (particles called K^+ and K^0 were expected to turn into lighter particles in a way which did not happen, even though Gell-Mann's three-quark theory predicted they ought to), but Glashow was largely going on a gut feeling that the universe should be pretty.

A lot of the time, scientists are hard-nosed sceptics who refuse to entertain any idea without evidence but sometimes they are human and they have hope.

Glashow believed nature was beautiful in a deep way and named his hoped-for particle the charm quark because its charming nature completes the quark symmetry. For his optimism, he was rewarded with its discovery in 1974. A reminder that sometimes in physics you can hope that maybe nature knows what she is doing.

THREE IS A MAGIC NUMBER . . . APPARENTLY

In the Arthur C. Clarke science-fiction masterpiece *Rendezvous with Rama* (1973), humanity discovers an abandoned alien structure built by a species obsessed with the number three. They find alien suits with three limbs, structures built in trios and every decision the mysterious species made seems to have been copied three times. Nature has a similar obsession.

The year before the charm quark was validated and, in fact, the same year *Rendezvous with Rama* was published, Makoto Kobayashi took the idea of symmetry and neatness one step further. QED had three matter particles – electrons, muons and tauons – so maybe a similar trend should be repeated in QCD.

The up quark's heavier sister was the charm quark and the down quark's was the strange. Might there be a third generation? Kobayashi, who did not believe in no-win scenarios, called these mega-quarks the 'bottom' and 'top' to complete the set. They were both discovered, in 1977 and 1995 respectively.

The humans in *Rendezvous with Rama* are left at the end of the novel with a lot of questions about who the aliens were and why they chose to do everything in threes. In particle physics, the story is the same.

Why are there three generations of quark and lepton? Nobody knows. Could there be a fourth of each kind? Nobody knows. Are the three generations somehow linked to the three colours? Nobody knows.

Maybe one day we will be able to overcome the gluon-thread confinement and analyse an isolated quark to get more insight into their behaviour. Maybe one day we will get our hands on a naked charm, a naked strange or a naked up quark. And maybe one day, if we are very lucky, we will catch a glimpse of a naked bottom.

CHAPTER FOURTEEN

Honey, Where's My Higgs?

HER MAJESTY, THE QUEEN OF PHYSICS

On 4 July 2012 newspapers around the world proclaimed a momentous day for science. The *Independent* front-page headline was 'Scientists prove existence of God particle', CBC news reported a 'missing cornerstone of particle physics' had been found and the *New York Times* ran with 'physicists find elusive particle seen as key to Universe'.

It was the monumental discovery of the Higgs boson and everyone was scrabbling to explain what the fuss was about. The Higgs boson is pretty complicated, however, and cannot be summed up in a newsworthy soundbite.

So complicated is the Higgs that in 1993 the UK science minister William Waldegrave offered a bottle of champagne to any scientist in the country who could summarise what the Higgs did on a single side of paper.[1] That is not what we are going to do here (I would rather have a milkshake) but we will try and get a feel for what the Higgs is all about.

What makes it important is that it validates a prediction made by physicists almost fifty years prior, requiring construction of the Large Hadron Collider – the largest machine ever built.

But how did we know it was worth the effort? I could invent a particle right now called the timon, the particle that makes you want to check your watch during a boring movie. Are we going to build a machine for that?

Come to think of it, how do theoretical physicists know *any* hypothesis is worth investigating? There are so many particles/fields and interactions between them, how do we figure out if our equations are sensible? Is there some ultimate law for making new physics laws?

The answer is yes and it comes from one of the most brilliant and criminally unheard of physicists of all time: Amalie 'Emmy' Noether.

At the beginning of the twentieth century, Noether was one of two women permitted to attend the University of Erlangen in Germany and had to get permission from every lecturer whose classes she wished to attend. Believe it or not, her uterus did not prevent her from understanding mathematics (how about that?) and she wrote a series of outstanding papers, which caught the attention of the well-respected mathematician David Hilbert.

Hilbert helped Noether get a job as a lecturer at Göttingen University, where she was the solitary female member of staff. The position was unpaid, of course, and her lectures had to be advertised under Hilbert's name but, regardless, she had a foot in the door of academia.[2]

What finally turned the tide was when she discovered what is probably the most important guiding principle in theoretical physics: Noether's theorem. In a way, it is a shame that to be treated as equal a woman had to triumph over every male physicist in the world, but also . . . it make her kind of badass. Men were not giving her enough respect, so she outmathed every one of them with a theorem so far-reaching it forms a cornerstone of QED and QCD, and solves puzzles in relativity even Einstein could not figure out.

Noether's theorem is about finding what physicists call 'symmetry', an idea with which we have vaguely been toying. When we are studying an event or a particle we use equations that tell us about the kinetic energy (movement) and the potential energy (location in a field). The difference between these two energies is called a Lagrangian and every physical law has one.

But we can always change the details of whatever scenario we are studying. If we perform our experiment near a strong magnet say, or change the mass of the particle, this will sometimes alter the Lagrangian and sometimes leave it untouched. If our changes do *not* change the Lagrangian, all the equations are identical and we call our theory symmetric. If changing something *does* impact the Lagrangian, the equations change too and our theory contains a 'broken symmetry'.

Noether's theorem says that if you have symmetry in a theory, there has to be an associated property of particles that does not change either.

For example, suppose we are studying a particle and decide to shift it 1 metre to the right. The particle behaviour will be identical. Our theory is therefore symmetric with respect to location.

Noether's theorem says this change in location is the result of the particle travelling with momentum from one place to the other, so momentum has to be 'conserved', i.e. it cannot be created or destroyed. Particles can transfer momentum during collisions but the total before and after is the same no matter what.

Another example of Noether's theorem is that when we move a particle forward through time the laws of physics are again invariant. Laws of physics are symmetric with respect to time so there must be a conserved property to go along with it – it turns out to be energy (since we are talking about cause/effect). Émilie du Châtelet had already proven energy is never created or destroyed but Noether's theorem gave us the underlying reason.

Charge is another conserved quantity, originating from the way a particle's wavefunction vibrates. That is why photons of light always create antimatter and matter particles in tandem. Charge has to be conserved so a chargeless photon cannot make an electron without making an anti-electron to keep the overall charge zero. And these are just a few examples.

Noether's theorem tells us what properties we can and cannot change for a law of physics and it was integral for Dirac, Feynman and Gell-Mann when figuring out how quantum field theories ought to work. Noether gave us a law for physics laws and it is hard to overstate how important that is.

Sadly, because Noether was Jewish, she was hounded out of Germany during the rise of Nazism and subsequently fled to America. But on the plus side she was received by a loving community of scientists who saw her as their undisputed queen. She earned her long-overdue recognition and her obituary in the *New York Times*, penned by Einstein, described her as 'the most significant mathematical genius thus far produced since the higher education of women began'.[3]

BE CALM, LITTLE ONE

One of the conserved quantities that emerge from Noether's theorem is called the lepton number. No rocket science needed to figure that one out: the number of leptons in the universe stays the same. It is a symmetrical law but irritatingly it looks *too* symmetrical on paper, because there is one known process where the symmetry is broken.

It is called beta decay, discovered by Madame Curie (the other queen of physics). It happens when a proton in the core of an unstable nucleus turns itself into a neutron, seemingly at random. As it does so, it spits out an electron, which darts away from the atom, detected by us as radioactivity.

In terms of a Noether law this makes sense because charge has to be conserved. If a neutral neutron turns into a positive proton it has to generate a negative electron too. But if lepton number has to be conserved alongside it, have we not just broken that law by generating an electron where there was none previously?

The explanation posed by Wolfgang Pauli (the guy who shot down de Broglie's pilot wave) was that there had to be another particle generated. Some sort of anti-lepton that possessed no charge.

Enrico Fermi called this hypothetical particle a neutrino, meaning 'little neutral one', and for twenty-five years we hunted them to try and prove Noether right. Unfortunately, neutrinos are the most inconspicuous and non-interacting particles known to physics so this was no simple feat.

Consider the neutrinos made in the core of our Sun where protons and neutrons change into each other all the time. A photon takes about ten thousand years to bounce its way to the solar surface from the centre, being absorbed and re-emitted by every particle it encounters along the way. A neutrino makes that same journey in twenty-three seconds.

Earth is constantly bombarded by solar-produced neutrinos and almost all of them shoot through the planet without blinking. Roughly sixty-five billion neutrinos passed through the tip of your little finger as you read this sentence.

It is not easy building a detector for something utterly uninterested in being detected. The world's biggest neutrino detector is the Super-Kamiokande (Super-K) near Hida, Japan, located 1 kilometre below the surface of a mountain (to filter out cosmic rays).

The Super-K houses a tank of 50,000 tonnes of ultra-pure water and trillions of neutrinos pass through it every second, most of them doing nothing. But every once in a while they strike an electron from its atom, which we can detect as a faint glow.

Neutrinos turned out to be real particles and thus lepton number is conserved. It took a quarter of a century to find them but they are elegant proof that Noether's theorem is correct. Oh, and of course neutrinos come in three generations called the electron-neutrino, muon-neutrino and tauon-neutrino.

A Sign of Weakness

The reason for the scarce interactions of neutrinos is that they have very few properties and do not couple to the fields with which we are familiar. They do not have colour, so will not talk to the gluon field, and they do not have charge, so will not talk to the electromagnetic/photon field either.

But neutrinos *will* occasionally interact with electrons. Plus, we know quarks emit them when changing identity from up to down, so there must be some field they are interacting with in order to do so. Something a lot weaker. It was named the weak field. Seriously.

Take an up quark with its $+\frac{2}{3}$ charge. When it turns into a down quark it becomes $-\frac{1}{3}$, meaning it loses +1 charge somehow. We have always thought charge stays put but the weak field might be violating that assumption.

The weak field could be carrying positive charge away from a quark in the form of a virtual positively charged 'weak particle'. Virtual particles never last, however, so this particle will decay soon afterwards, transferring energy into the positron and neutrino fields to conserve charge and lepton number. We can explain an up quark turning into a down quark like this:

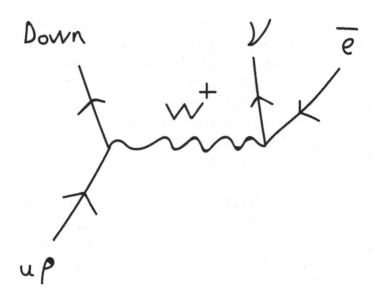

Read from bottom to top, we begin with an up quark. It couples to the weak field and generates a positive weak particle (W⁺), turning into a down quark itself.

This positively charged weak quantum then decays, making a regular neutrino (represented by that wonky V symbol) and a positron to conserve the positive charge (shown as an electron with a little bar above it). The reverse process also happens, except with a W⁻ particle.

You are probably thinking that the weak-field particle must have some awesome name to go with photon and gluon but I'm afraid by this point everyone seems to have got bored, so they tragically called them W particles. W⁺ if carrying positive and W⁻ if carrying negative.

The property a particle needs to couple with the weak field and emit W particles is called 'weak isospin' and comes in two varieties, +½ and −½. Quarks, leptons and neutrinos have weak isospin but its coupling constant is very low, so we rarely see its influence.

But what in the name of Erwin Schrödinger's cat-munching ghost happens when neutrinos meet? They both have weak isospin, which means they should create virtual particles between them. This cannot happen through W⁺ or W⁻ interaction because neutrinos are chargeless. There has to be a third weak particle available, one with no charge.

Sheldon Glashow named it the Z particle, because it stands for zero charge, I guess? The Z and W particles were observed in 1973 and 1983 respectively at the Gargamelle detector in Switzerland (named after the giant from the François Rabelais novel *The Life of Gargantua and of Pantagruel*, not the incompetent bad guy from *The Smurfs*).

The Z and W discoveries verified the behaviour of neutrinos, confirmed the existence of the weak field and once again proved Noether's symmetry laws correct. But as you probably know by now, quantum physics is like an infernal Rubik's Cube. As soon as we solve one part we muddle up something else.

A Totally Useless Idea

All quantum field theories involve two types of object: matter particles (quarks, electrons, neutrinos) and interacting force-field particles (photons, gluons and W/Zs).

Matter particles are collectively called fermions and have properties such as occupying space, whereas force-carrying particles are collectively called bosons and are able to overlap.

Your body is made of fermions (electrons and quarks), which is why you take up a volume. A beam of light, on the other hand, is made from bosons (photons specifically), which is why torch beams pass through each other rather than banging together like lightsabres. In that respect, bosons are a disappointment.

When a particle interacts with the weak field it often involves a change in charge, so the weak and electromagnetic fields are obviously coupled. The early quantum field theory for the weak field was called 'quantum flavour-dynamics' (QFD) but since the weak and electromagnetic fields talk to each other the full theory, which includes all photon and weak interactions, is called 'electroweak theory' and won a Nobel Prize for Steven Weinberg, Abdus Salam and Sheldon Glashow. I would have called it quantum electroweak-dynamics myself because it would give us the acronym: QEWD.

It is a very symmetric theory but this elegance is also its biggest flaw since the photon and weak fields are drastically different.

Ws and Zs are short range but photons travel forever. The weak force requires three different particles while the electromagnetic force works with only one. Then there is the biggest broken symmetry of all: W and Z particles have mass and photons do not. That is not something we expect for force-carrying bosons that overlap, so something was breaking symmetry between the photon field and the weak one.

In the mid-1960s three people independently hit on the same solution. Robert Brout, François Englert and Peter Higgs suggested you could preserve electroweak symmetry the same way Pauli solved the lepton symmetry problem – add a new field/particle.

At the start of the universe, they argued, in the initial flickering moments of creation, the electromagnetic and weak fields were identical. But there was a third field lurking in the background and when this field 'switched on' everything changed.

This field is not like the others because its resting value is not zero; it is an actual number everywhere. Because of such an unusual property this new field had the ability to create different particles out of itself rather than one. These quanta are called Goldstone bosons, after physicist Jeffrey Goldstone, and they couple to the weak field majorly.

Prior to this field sticking its nose in, particles of the weak field were massless and infinite ranging, just like photons, but when the Goldstone bosons mixed with them, their character changed and they became the W^+, W^- and Z.

I like to imagine the weak field laid carefully on top of this eccentric new field like wallpaper placed on a wall. The wall has three kinds of bump on its surface (Goldstone bosons) and they push through the weak wallpaper, making it look like three kinds of weak particle to any casual observer.

Since the photon field does not couple to this additional field its particles are left untouched, becoming the photons we are used to. Boom. Symmetry successfully broken . . . assuming you could detect this freaky field. Which, no surprise, you cannot.

The idea of a non-zero field 'switching on' at the dawn of time and having three types of particle was not only exotic, it was untestable since Goldstone bosons are hidden within weak particles, making them invisible.

The journals to which Brout, Englert and Higgs submitted their idea all refused the idea, with one even replying that it was 'of no obvious relevance to physics'.[4] It was neat mathematically, but you could not test it, which prompted Higgs to grumble to a member of his research team, 'this Summer I have found something that is totally useless'.[5]

Broken Mirrors

We know the story has a happy ending though, so we should look at this new field closer. In some of the literature you see references to the weak-field particles 'eating' Goldstone bosons. How does this happen? The answer is to do with a particle property we have ignored until now because it, too, looks totally useless.

Charge, colour, spin, weak isospin, etc. determine how particles interact with various fields but Dirac's quantum field theory predicts another property called chirality, from the Greek word for 'handedness'. It is a property particles have and it does absolutely nothing.

Chirality can be described mathematically but has no obvious physical meaning. We just know that each particle/field seems to oscillate from one chirality to another, back and forth like the tick-tock of a clock, and we call the two chiralities left- and right-handed. It does not mean particles are waving alternating little hands into the air like some dad-disco move, although they might as well be. We have no idea.

For most of quantum history, chirality sat there in the equations hopping from left to right doing plum nothing. Particles flip their chirality to one handedness, then flip back. Then they flip again. And then back. Until you introduce the weak field.

In 1956 Chien-Shiung Wu and her team were conducting experiments on cobalt atoms to test how symmetric the weak force was. The strong and electromagnetic forces work the same whichever way a particle is facing but Wu discovered that the weak force breaks symmetry. Radioactive decay particles are only emitted from an atom whose chirality is in the left-hand state.

Although a particle like a quark has weak isospin (the weak field property) all the time, its ability to interact with the weak field is only present in its left-hand chirality. We say that a left-handed particle has a 'weak hypercharge of +1' and a right-handed particle has a 'weak hypercharge of 0', i.e. it will couple to the weak field or not.

At this point you start to feel like the whole of quantum physics is a loopy bunch of properties assigned at random during the big bang with no rhyme or reason. But if you want the laws of physics to be a little more orderly I am afraid you need to find another universe. I recommend the one where I am Batman. Unless you are already in that one, in which case stay put.

Chew Your Bosons with Your Mouth Closed

Let us go back to the Z boson, a particle with no colour or charge. It is essentially a photon with two differences: it has mass and it couples to the weak field when in its left-handed chiral state.

Because the Z boson is alternating between coupling to the weak field and not, its weak hypercharge is going back and forth from +1 to 0 over and over. But look out, who is that coming over the hill towards us? Why, it's Emmy Noether!

'Guess what?' she says. 'Weak hypercharge is a conserved property too.' And with that she drifts away, smiling contentedly to herself at the damage she hath wrought.

According to Noether's theorem, if the Z boson is switching its weak hypercharge on and off, that property must be going to and coming from somewhere. It is a conserved property so it cannot vanish or appear. There must be a field that can absorb and donate weak hypercharge from/to the Z boson. And this is what the Brout–Englert–Higgs field does.

Each time the Z boson becomes left-handed it sucks up weak hypercharge from the field (absorbs a Goldstone boson) and when it flips back to right-handed chirality it returns weak hypercharge to the field (releases a Goldstone boson).

I imagine a Z particle as being like one of those chattering-teeth toys that are constantly opening and closing, accepting and spitting out weak hypercharge as they move.

In a Feynman diagram this means a Z particle is mixing and un-mixing with the Brout–Englert–Higgs field constantly, so when we discovered a Z particle we were discovering the Goldstone boson 'in its mouth' as well.

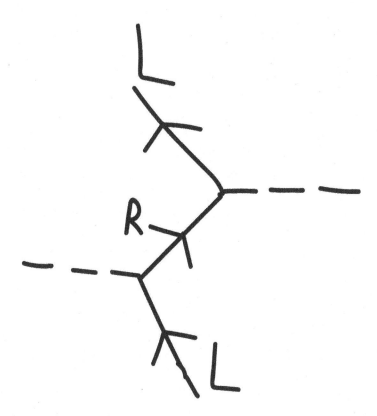

This Feynman diagram shows a Z particle zigzagging its way up the page, hopping chirality from left to right. It starts off in the L state and then flips to the R, releasing a Goldstone boson in the process, which carries away the weak hypercharge (shown as a dashed line). It travels in the R state for a moment but then flips back to the L state, absorbing a Goldstone boson and regaining weak hypercharge. Forever and ever and ever.

WHO CARES?

When two particles interact, their fields couple. But if a particle is changing its identity constantly the coupling can be lost.

It is as if a Z particle changes mood every few seconds, switching from jokey and cheerful to sombre and serious. If you want to tell the Z particle a joke you have to be quick because it will flip to deadpan momentarily and your interaction will be lost. Or if you want to tell the Z particle about the death of your pet horse you need to be equally quick because in a moment it's going to start cracking dead-horse puns in your face.

As the Z particle flips its personality, it becomes harder to interact with, and a particle which is hard to interact with is not easily influenced. Z particles will barrel through the fields around them like a bullet through fog, but a particle that does not change personality is easier to affect.

In other words, rapid chirality-flipping makes a particle hold its trajectory whereas slow chirality-flipping makes a particle easier to deflect. We have just described the difference between something being heavy and something being light.

A photon can be bounced around because it is massless, but a Z particle is not easy to grip onto and holds its momentum, making it very heavy. Chirality-flipping turns out to be where mass itself comes from so the Brout–Englert–Higgs field allows the Z particle to have mass and the photon to be massless!

It is not just the Z particle either. Something like a muon is flipping its chirality at a faster rate than an electron, meaning it would be even harder to slow down and thus would present as a heavier particle. Which of course, it does.

In fact, the W^+, W^-, electrons, muons, tauons and all the quarks are flipping their chirality constantly, which means their hypercharge needs to be conserved through the Brout–Englert–Higgs field. Each particle couples to this field via a mechanism called Yukawa coupling and turns out to be the very reason particles are able to have mass in the first place. Not so useless now, eh?

HOLY MASS-CONSERVING SYMMETRY-BREAKING FIELD, BATMAN!

As we have already seen, there is no way of detecting the Brout–Englert–Higgs field. The Goldstone bosons which allow particles to chiral-flip (gain mass) are mixed in with the particles themselves and we cannot get them on their own.

In fact, the Nobel laureate Leon Lederman wrote a whole book about this frustration. Originally he wanted to call it *The God-damn Particle* because Goldstone bosons are so god-damn hard to find, but his publishers were not happy with that so he shortened it to *The God Particle*. Far less controversial.[6]

So, how do you detect a field whose particles are always mixed in with others? This is where Peter Higgs went further than Brout and Englert, and why people started referring to the whole thing as the Higgs field (probably also because it is easier to say).

Since this field is different to the others, Higgs believed it could do something novel: carry shockwaves. The other fields have zero value at every point in space but the Higgs is non-zero and if you gave it enough of a whack you can send a momentary compression through it, detectable as a momentary quantum in the Higgs field. A Higgs boson.

Technically the Higgs boson does not do anything interesting, it just proves the Higgs field is there, which is why physicists had difficulty explaining it in 2012. The Large Hadron Collider has a circumference of 27 kilometres, came at a price tag of $9 billion and costs a further $1 billion a year to run, drawing 1.3 terawatts of electrical power. If a journalist asked the very fair question, 'What does the Higgs boson do?' it would be pretty unwise to answer by saying, 'Oh, it does nothing.'

It is really the Higgs field and the Goldstone bosons that do interesting stuff, but they are slippery rascals so we have to create a pointless particle in the field to see if it is there. The Higgs boson does not give particles mass but it proves our theory of how they get mass is correct, and that is what the Large Hadron Collider was all about.

You take a bunch of hadrons (particles made from quarks) and send them into a giant tube, which uses electromagnetism to accelerate them around a loop, like an epic centrifuge. Cooled to a few degrees below the coldness of space, moving at 99.9 per cent of the speed of light, these hadrons are then smashed together at various points around the ring and the energy released ends up distributed into all the fields.

You sometimes hear it phrased as if quarks are smashed together and particles inside come falling out, but that is not right. Quarks do not have particles inside but when we slam them together, the energy released is so enormous it gets transferred to every other field and the collision creates electrons, muons, tauons, neutrinos, antimatter, gluons, photons, Ws, Zs and, if we are very lucky, it will shake the Higgs field and we will see a tiny blip on our readout.

The momentous announcement was finally made to a packed lecture hall in France on Independence Day 2012, using a PowerPoint presentation written in comic sans, no less. A particle with the expected properties of a Higgs boson had at last been found.

Peter Higgs, who was in the audience, began weeping as applause detonated around him. His forty-eight-year search was over and his hypothesis was vindicated.

THE HOUSE THAT EMMY BUILT

The complete list of particles in quantum field theory is huge. Take the number of quarks, for instance. We start with the six main fields: up, down, charm, strange, top and bottom, but each one can also be found in three possible colours (red, green and blue), giving us eighteen particles/fields. Then we have to include all the antimatter quarks, giving us thirty-six. And if you want to include the left and right chiral versions, that becomes seventy-two.

Most of the variations are nearly identical, however, so rather than putting every possible version of a particle into a periodic-table of particles we can draw up a simplified list such as this:

FERMIONS (MATTER)			BOSONS (FORCES)
Up Quark	Charm Quark	Top Quark	Photon Gluon × 8
			W⁺
Down Quark	Strange Quark	Bottom Quark	W⁻ Z
			Higgs
Electron	Muon	Tauon	
Electron Neutrino	Muon Neutrino	Tauon Neutrino	

What you are looking at here represents a century of scientific experiment tied together with the symmetric bow of Noether's theorem.

The foundations she laid were built upon by Dirac, Feynman, Gell-Mann, Weinberg, Salam, Glashow, Higgs and many others to create a structure of ingenuity and beautiful complexity.

It is called the standard model of particle physics because it models every particle and every interaction between them and is the crowning achievement of physics.

To some, the Large Hadron Collider might seem like a waste of effort but it is worth remembering this was a machine built to test the grandest theory we have about the universe. Quantum field theory and Noether's theorem present us with the biggest question, so it makes sense that our response should be to build the biggest answer.

CHAPTER FIFTEEN

The Trouble with G

Everything Ever Almost

The reality we live in owes almost all its beauty and complexity to the three forces of quantum field theory: the strong, the electromagnetic and the weak.

Without the strong force, nuclei at the centres of atoms would not be stable and protons would simply fly apart before anything could form. The gluons and quarks that bind everything together are the reason we are treated to a universe of more than just hydrogen. The carbon, nitrogen, oxygen, phosphorus, sulfur, sodium, chlorine, calcium and iron in your body, in fact the entire periodic table of elements, owe their existence to the laws of quantum chromodynamics.

Those elements, when combined, can form the intricate molecules of chemistry with electrons exchanging energy through photon transfers, making every reaction possible. Not only that, the ground you stand on which repels against you is only doing so because electrons in your feet and electrons in the Earth are able to exert a force on each other due to the interplay of electrons and photons.

When you push or pull against something, when you feel friction or drag, when you experience buoyancy, upthrust or any other phenomenon described by Newtonian mechanics, you are experiencing electric repulsion from the photon field, not to mention the fact you can see any of this in the first place due to the existence of light. All of chemistry, all of classical physics and all of optics owe their behaviour to the laws of quantum electrodynamics.

And then there is the magnificence of biology. The most complicated of the sciences whose simplest structures are made from molecular chains billions of atoms long. As these DNA strands in your cells replicate, mutations in the code lead to new features which get passed along, allowing a species to diverge and evolve over time. These mutations are sometimes caused by transcription errors in the copying process, but that only takes things so far. The real gift to life on Earth are the radioactive particles that filter into our atmosphere and disturb the coding process: remnants of solar winds and astronomical phenomena caused by particle decays out in the coldness of space.

Without W or Z bosons, such radioactive decays would not occur and life on Earth would be confined to a few dozen bacteria swirling about in rock pools. And without the Ws or Zs, quarks in the core of the Sun would not be able to turn protons into neutrons, preventing nuclear fusion from taking place and thus preventing the Sun from shining at all. The overwhelming splendour of biodiversity and the sustainability of life itself is only possible due to the laws of electroweak theory.

The quarks and leptons that make up your body, the photons and gluons, the bosons that tell them to interact, the Higgs that give you mass and the neutrinos that keep everything balanced, are all accounted for in the standard model of particle physics and the underlying quantum field theories.

We do not know everything, and there are still wonderfully juicy questions to be answered but quantum field theory points us in the right direction. We are taking baby steps in this new realm, it is true, but we are no longer fumbling blindly. Every event in the history of history is the result of particles interacting through fields and we now have a framework to explain them all. Except for one problem. The thing quantum field theories cannot handle. Gravity.

WEAKER THAN WEAK

The story of the apple striking Newton's head is slightly apocryphal. He was actually watching it fall from a tree near his home in Lincolnshire, when it struck him as odd.[1]

Most of the forces on a falling apple can be explained using simple mechanics. The snapping of the branch is the result of atoms in the stem rearranging. When it moves through the air it 'bumps' into air particles, slowing its acceleration, and once it hits the ground we can use more simple laws to explain why it lands, rolls and splits in a certain way.

All of these phenomena are the results of particle–particle interactions, mostly handled by QED and electron-photon theories. But what makes the apple fall in the first place? That is the real question.

Newton did not invent gravity; sadly, things were not floating about in the air prior to 1687. What he realised was that gravity is a force of its own. Apples do not move downward out of preference as had been supposed by Aristotle; they get faster during the fall, which means something is actively pulling on them.

A simple way of proving this is to show that an apple hitting the ground from a tree hits with more of an impact than one being dropped from a few centimetres up. Clearly the longer you are falling for, the faster you get, and if something is making the apple get faster, something must be exerting a force.

Newton's realisation was that objects with mass (later modified by Einstein to include energy) are in communication with each other through some invisible attracting medium we now call the gravitational field. Along with the weak, electromagnetic and strong fields, gravity is a fundamental force of nature.

Your gut instinct might be that gravity is the strongest of the four, but in fact it is weaker than the weak force by a factor of about a trillion. The reason gravity is still a force to be reckoned with (you're welcome) is because it has infinite range and acts on absolutely *everything*. All the quarks, leptons, gluons, photons, Ws, Zs and the Higgs are drawn together by its ruthless greed.

As the strong, weak and electromagnetic forces are jumping merrily from particle to particle, barely noticing gravity, it is lurking in the shadows, subtly and silently binding things together without remorse, and nothing escapes it. Not even light. Not even time itself.

Right now, as you sit reading this, the book in your hand is gravitationally attracted to your face and vice versa. You do not notice this amount of gravity because you need a planet's worth of the stuff for it to become important, but everything around you is collapsing towards everything else very, very slowly.

You can defeat gravity temporarily and on a small scale – a magnet can lift a paper clip off the ground with ease – but when you look at the big picture, gravity is holding everything to the Earth's surface and will not let go without a wrenching effort.

The other three forces can be explained by introducing virtual particles moving between matter and so an analogous particle can be introduced to account for gravity: the graviton.

Discovering the graviton is going to be tricky though, because it is such a weak force that in order to work up enough energy to agitate the field, we would need a particle collider roughly the size of the galaxy. And even if we did that, the energy of such a collision would be so tremendous it would make a black hole at the point of impact, which would suck the gravitons back inward and we would never see them. Unless we can figure out another way of detecting particles and fields, gravitons are going to remain hidden.

Odd One Out

Gravity is different to the other three forces in many ways. It is not just a little bit asymmetric from them (the way electromagnetic and weak are a bit asymmetric from each other), it is the undisputed Quasimodo of physics. Here is some of the nastiness we run into when we talk about gravity.

1. Gravity is weaker than the other forces by an alarming amount. If we represent the strengths of the other three forces like pins stuck on a line, we could place them all within a centimetre of each other as they are all comparable in strength. Gravity's pin would be somewhere out towards the Andromeda galaxy.

2. Using quantum field theory we can predict how much energy there should be in the vacuum of empty space (the result of adding all those virtual particles together). The total amount of energy in the vacuum should be around 10^{105} Joules per cubic centimetre but when we actually measure the energy for real,

which we do by observing the impact of gravity on galaxies, we get 10^{-15} Joules per cubic centimetre. The value quantum field theory predicts and the value gravity measures are out by several million quadrillion. Quantum field theory on its own boasts the most accurate prediction in all of science, as we have seen, but including gravity in the equation gives us the *worst* prediction in science.

3. One of the features of fermions (matter particles such as quarks, electrons and neutrinos) is that they occupy their own space. It is a feature called the Pauli exclusion principle, which states that particle identities including their energies and locations are exclusive to them. Fermions stay separated at all costs, but when we get enough gravity together at the heart of a black hole, particles get crunched together. In quantum field theories, the Pauli exclusion principle is inviolable. In gravity theories, it can be broken quite easily.

4. The theory that describes how gravity works is Einstein's general theory of relativity, which relates energy, mass, time, light and empty space. The official record states that Einstein came up with it in 1916 although what is less widely known is that he originally discovered it in 1912 but threw the equation away and told nobody,[2] presumably because he thought it was wrong.

   General relativity is a theory which describes empty space bending around objects to create distortions, which we perceive as gravity. The theory matches experiment perfectly and rests on a key assumption: space is smooth at every point with a clearly defined value. In quantum mechanics, however, the Heisenberg uncertainty principle says that can never happen. All fields and particles jiggle, which means no theory describing things as smooth can possibly be correct. Gravitons themselves would have to obey Heisenberg uncertainty laws while gravity itself would apparently not.

5. The other three force fields are sitting against the backdrop of empty space. Particles can be bumped out of them but the shape of the fields obeys sensible geometry. In general relativity, however, empty space can curve.

All the particles used to moving in straight lines suddenly find the space around them changing shape and we have no way of explaining how this affects them. We can use gravity to explain what happens to a whole crowd of particles, but the force of gravity on a single particle cannot easily be calculated.

Gravity is not just the loner kid at the house party. It is the bully who goes around kicking everyone in the back, knocking over lampshades and urinating on the television. Quantum field theory works neatly until we factor in gravity, at which point it all breaks down. And this is the most thrilling thing that could have happened.

The Tree of Knowledge

Every object near the surface of the Earth is pulled towards it through gravity. Newton realised that this same force was working on all the suns, moons and planets in the cosmos because gravitation was a universal law, applying to the heavens and the Earth alike. Thanks to him, two apparently different domains of reality were connected by one simple explanation.

A few centuries later, Einstein discovered that the laws of energy were part of the same framework and incorporated them into relativity. Once again showing that two separate branches of physics were bound together by a previously hidden connection.

Meanwhile, Michael Faraday discovered that electric and magnetic fields, thought to be separate, were different facets of the same field, giving us a unified explanation that encompassed electricity, magnetism and light.

Then quantum physicists took these electromagnetic laws and joined them with particle physics to birth QED, before combining QED with radioactivity and the weak force to give us electroweak theory.

The same thing happens again and again. We start at different places in our knowledge, places that appear unrelated, but as we follow them to their logical conclusions we discover links between unconnected theories like twigs joining up to form the branches of a tree. The more we study the universe, the more those branches connect.

At the moment, our tree of knowledge is a network of theories that sprout from three main boughs. One is general relativity, which explains astronomy, cosmology and gravity. One is the electroweak theory, which explains mass, light, radioactivity, classical forces and chemistry. The third is quantum chromodynamics, which explains the nucleus. And we are pretty close to uniting those last two.

Right now we are finding ingenious ways to combine the electroweak theory with quantum chromodynamics to give us a 'grand unified theory' or GUT that will account for the entire standard model in one go.

If we can devise a successful GUT, we would have everything in physics explained by two boughs: general relativity for gravity and the completed quantum field theory for everything else. Could we eventually bring them together to form a trunk? Is there a single theory of everything that will explain away the contradictions between gravity and the quantum forces? We do not know but we are sure as hell going to try.

THE BEGINNING

A century ago we thought we had the answers. Quantum physics has taught us some humility since then. Rather than science drawing to a denouement, it appears that things are just getting started and that is a good reason to get excited. We have a daunting task ahead of us but the fact that we have got so far in such a short time gives me tremendous hope.

Humans are born with not only a thirst for knowledge but a brain capable of acquiring it. We do not like the phrase 'nobody knows', and we are determined to find out where we fit in the grandest scheme. That is why we never give up answering questions and questioning answers. The universe may be unimaginably complicated, but if we can make sense of quantum mechanics, who knows what else we can do?

For that reason, and for many others, I truly believe that science will save our species.

Timeline of Quantum and Particle Physics

Science rarely happens in straight lines. Sometimes the theory we invent makes a prediction we accidentally verified years before, without realising what we were doing. From our current vantage point we can put the puzzle pieces together in a narrative, but we sometimes lose historical accuracy in the process.

Throughout this book I have focused on storytelling in order to make this complicated topic easier, but I have occasionally sacrificed chronology to do so. Here, in the interests of authenticity, is a more precise timeline of how things went down.

1618   Descartes proposes light to be waves in the plenum.

1672   Newton proposes light to be made of corpuscles.

1801   Young does the double-slit experiment showing light is made of waves.

1846   Faraday speculates light is an electromagnetic wave.

1861   Maxwell proves him right.

1897   J. J. Thomson discovers the electron.

1899   Rutherford discovers that radioactivity is made of particles.

1900   Planck invents light quanta.

1905   Einstein proves all matter is made from atoms *and* that light is made of photons *and* publishes the theory of special relativity. Buys cake on birthday.

1908   Rutherford discovers the nucleus.

1912   Einstein discovers general relativity. Tells no one.

1913   Bohr discovers electron energy is quantised in shells.

1915   Noether comes up with her theorem. Girl power prevails.

1916   Einstein rediscovers general relativity. More open about it.

1917   Rutherford discovers the proton.

1922   The Stern–Gerlach experiment is conducted. Makes no sense.

1924   De Broglie suggests wave–particle duality.

1926   Schrödinger writes his wave equation.

1926   Born interprets the wavefunction as the square root of probable behaviour and properties.

1927   Pauli adapts the Schrödinger equation to include 'spin'.

1927   Heisenberg discovers the uncertainty principle.

1927   George Thomson shows electrons can be diffracted like waves.

1927   De Broglie presents the pilot-wave interpretation.

1928   Dirac comes up with quantum field theory.

1930   Heisenberg outlines the Copenhagen interpretation. Einstein is not happy.

1930   Pauli proposes the existence of neutrinos.

1932   Chadwick discovers the neutron.

1932   Von Neumann tries to find the source of wavefunction collapse. Finds nothing.

1932   Anderson discovers the positron.

1933   Fermi proposes the weak field.

1935   Schrödinger suggests we kill/do not kill a cat.

1935   Yukawa proposes the strong force to explain nuclear stability.

1935   Einstein, Podolsky and Rosen publish a paradox.

1936   The muon is discovered.

1939   Batman is born.

1947   The pion is discovered.

1947   The kaon is discovered, acting strangely.

1949   Feynman, Schwinger and Tomonaga create a successful form of QED.

1952   Bohm expands on the pilot-wave interpretation.

1956   Electron neutrinos are finally discovered.

1956   Wu discovers the weak field is asymmetric with respect to chirality (and therefore weak hypercharge).

1957   Everett proposes the many worlds interpretation.

1961   Wigner suggests consciousness could trigger wavefunction collapse.

1962   The muon neutrino is discovered.

1964 Bell proposes a way to test the EPR paradox.

1964 Gell-Mann outlines quantum chromodynamics including up, down and strange quarks.

1964 Glashow proposes the charm quark . . . because obviously.

1964 Brout, Englert and Higgs propose a new field to explain mass.

1968 The up, down and strange quarks are discovered.

1968 Weinberg, Salam and Glashow complete the electroweak theory.

1971 Hafele, Keating and Mr Clock verify relativity.

1973 Kobayashi proposes the top and bottom quarks.

1973 The Z boson is discovered.

1974 The charm quark is discovered.

1974 The tauon is discovered.

1974 The tauon neutrino is discovered.

1977 The bottom quark is discovered.

1982 Aspect successfully carries out a Bell experiment, proving classical physics cannot explain entanglement.

1983 The W^+ and W^- bosons are discovered.

1986 Cramer proposes the transactional interpretation.

1993 Peres, Wootters and Bennett propose quantum teleportation.

1994 Tonomura carries out the single-electron double-slit experiment, proving unambiguously that particles self-interfere.

1995 The top quark is discovered.

1998 Construction begins on the Large Hadron Collider.

1999 Kim builds the first delayed choice quantum eraser, showing apparent backward-time quantum entanglement. Presumably sends message to Cramer in 1986.

2005 Couder gives some evidence that might validate the de Broglie–Bohm interpretation.

2008 Large Hadron Collider switches on for the first time.

2012 Large Hadron Collider discovers the Higgs boson.

2014 O'Connell puts first classical object in quantum superposition.

2015 Bohr potentially rules out the de Broglie–Bohm explanation.

2017 Jianwei achieves record quantum teleportation to a satellite.

2017 Lidzey accidentally entangles some bacteria with a laser beam.

2018 Vanner creates a quantum drum.

A Closer Look at Spin

Spin comes in quantised multiples of a number called Planck's constant, which is the number you get if you divide the energy of a particle by its associated frequency and it always comes to the same value: 6.6×10^{-34} Joule seconds. Particles can have spin values that are either half or whole multiples of this number, i.e. a particle can have a spin value of ½ Planck's constant, 1 Planck's constant, 1½ Planck's constant, 2 Planck's constant, 2½ Planck's constant and so on, as well as being positive or negative.

Particles with half-value spins are called 'fermions' while particles with full-value spins are called 'bosons' and they behave very differently (examples of this are in Chapter Fourteen). But not all particles are magnetic, despite them all having spin.

There is a term for a particle's magnetic character: its 'magnetic moment'. Magnetic moment determines how strong a particle's magnetic field is and it arises from the following relationship:

$$\mu = g \, \frac{e}{2Mc} \, S$$

The symbol μ is the magnetic moment of the particle, which we can think of as 'magnetic charge'. The g is called the gyromagnetic ratio and is a number unique to every particle, relating the other properties together.

The e represents electric charge, the M represents mass, c represents the universal speed limit (see Chapter Eight) and S represents something called a spin matrix, which is a 2×2 number grid keeping track of the different ways a particle can be spinning.

For an everyday object we can define spin using something called 'angular momentum', which measures how heavy the particle is, how fast it is spinning and which way round the rotations are happening (clockwise or anticlockwise). For quantum mechanical spin these numbers are not enough and we have to describe it as having four possible ways of pointing (we call them the four vectors of spin). Sometimes spin is referred to as 'intrinsic angular momentum' because it is a property that resembles angular momentum but it is nestled within the particle's identity even when it is stationary.

What this equation reveals is that the magnetic moment of a particle is a product of all its properties together. In the Stern–Gerlach experiment, what they were actually measuring was the magnetic moment of silver atoms but since the mass, charge and g were identical for each particle, the two directions in which the atoms flew had to be a result of S – the spin property. So although it was not quite measuring the spin of each particle (we have no way of doing that directly since we do not know what a spin experiment would even look like) for all intents and purposes, magnetism lets us measure spin differences.

What is also important to note is that in order for a particle to be magnetic it must have both spin and electrical charge. A particle that has spin but no charge, e.g. a neutrino (a particle discussed in Chapter Fourteen), has spin of ½ but no electric charge, so in the equation above we would write zero for the e term and the overall answer would be zero as well. Electric charge and magnetism are always linked, and if a particle has one it definitely has the other.

Solving Schrödinger

Solving the Schrödinger equation for a single electron around a single proton (a hydrogen atom) is doable. But when you include more particles it becomes very tricky, very fast.

A helium atom has two protons and two electrons so you need to include interactions from both electrons to one proton, both electrons to the other proton, both electrons to each other, both protons to each other, and then combine them all. In three dimensions.

The bigger our atoms or molecules get, the more interactions we have to handle and it reaches the point where even powerful computers struggle to take everything into account. It therefore makes sense to use a few approximations in your sum, which saves time while still giving answers close to the full-on Schrödinger version.

One of the approximations we often make is called the orbital approximation. You imagine your atom has only one electron and you boost it to a higher energy, forcing it into all the high-energy orbitals.

When we try to work out the shape of an atom with twenty-six electrons, for example, we often do it by imagining a hydrogen atom and boost its electron twenty-six times until it looks about right.

The result is like a child doing an impression of an adult by standing on stilts and wearing bigger clothes. It is not quite an accurate picture but it is a decent way of getting a feel for 'this is what a bigger version might look like'.

Another technique we can use is called the Born–Oppenheimer approximation, where we make the assumption that the energy and vibrations of the nucleus are so slow compared to those of the electron that they can be ignored. We imagine the electrons are orbiting/waving around a single positive point, which does not have an interesting life of its own. This allows us to focus on electrons and their interactions exclusively, without worrying about how the nucleus is going to interfere.

Hands down though, the best way of fudging the Schrödinger equation is a method called density functional theory, invented by Walter Kohn and John Pople, who shared the Nobel Prize for its invention in 1998.

Density functional theory, or DFT to those dwelling inside the circle of nerds, is a beautiful way of solving a molecular wavefunction with loads of particles in it. Rather than modelling each particle as an individual point and calculating every interaction one at a time, DFT replaces it all with an 'electron cloud' representing all the electrons in one big smush.

Once you have blurred every electron together and calculated the 'thickness' of the electron density you can talk about how the atom or molecule behaves over time. Where the cloud is thickest corresponds to where the electrons are most likely to be and where it is thinnest is where the electrons are rarely observed.

A DFT calculation on a small molecule can be processed in a few hours, giving you an answer that is usually over 90 per cent accurate. Compared with solving the Schrödinger equation (which would take years for a large molecule) it has become the industry standard for quantum calculations.

Einstein's Bicycle

This simple exercise, which illustrates light-speed permanence, comes from the science author and television presenter Carl Sagan. He imagined a scenario where a cyclist is pedalling towards you down a road, when suddenly a large truck cuts across the cyclist's path and they swerve to avoid it.

The truck is not advancing towards you, so the light coming off its flank is approaching you at regular light speed, which physicists represent using the letter c for *constant*. The cyclist *is* pedalling towards you, however, so the light coming from them approaches you at c + their cycling speed. This means any light from the bicycle will reach your eyes before any light from the truck.

When the truck cuts in front of the cyclist, the cyclist swerves to one side and the light coming from their new position (telling you that they moved) will reach you first, followed soon after by light from the truck as it cuts across the road.

What you should see is the cyclist swerving for no apparent reason (the light from the truck has not reached you yet) and then a few seconds later the truck pulling out behind it. You would, in this scenario, wonder why the cyclist swerved a few seconds too early. But of course that is not what happens.

The light coming from the truck and the swerving bicycle hit your eyes at the same, telling a more sensible story. But, since the bicycle was moving at a faster speed, the light beam it sent out should have arrived first. The only way of accounting for this is that the speed of light coming from the bicycle was not c + cycling speed, but c itself, the same as the truck. Light speed must therefore be the same, no matter how fast everyone is moving.

Taming Infinity

A lot of the problems in theoretical physics come from infinity. Take the double-slit experiment, for instance. We can pierce two holes in a wall and send a photon towards it, calculating where it is likely to arrive by combining the two possible paths as probable outcomes.

If you cut a third hole in the wall, the story is much the same. You calculate three routes for the photon instead of two and compute the probable outcomes of all three. Same thing with four holes, forty holes or four hundred. But eventually you get to a point where the wall has so many holes it is no longer a wall but a great big empty space.

This means we have to calculate an infinite number of paths for the photon because there is an infinite number of holes (no wall = infinite holes). Yet, if we shine a photon at a detector screen with no wall in the way, the photon obviously goes in a straight line. It would appear that the photon 'sniffs out' (Feynman's phrase) the infinite possible paths it can take and then picks the classical route as if the infinities somehow cancel out.

Another can of worms in QED is the issue of particle self-interaction. An electron has a negative charge, which means it interacts with other negatively charged particles. Technically, an electron should therefore interact with *itself* seeing as it is charged, but because an electron is infinitely close to itself, the self-interactions give infinite answers when we try to compute them.

These two examples are a real pain because infinity is not a real thing in physics. It exists in the world of abstract mathematics but in the actual universe there are no infinities (the universe could not fit them), so when a theory predicts an infinite answer that is a sure sign something is wrong with the theory.

When an equation starts heading toward infinity, scientists say it is 'blowing up' and a lot of theoretical physics is spent trying to defuse these numerical explosions. Usually by altering the equations, deriving new ones, or changing the input values to get more sensible answers.

One of the clunkier tricks is to simply chop the numbers off at the point where they get too big (a method called regularisation) but this is hopelessly crude; not much better than crying about the equation and ignoring it.

A more sophisticated approach is to do something called 'renormalisation'. The idea this time is to pick properties for the fields (mostly from educated guesswork) and solve a bunch of different equations with these values until you get matching answers. The more details you include, the closer you get to experimental results.

It is the mathematical equivalent of compositing a sketch of a criminal from a number of eyewitnesses. You start by making a few assumptions, e.g. the facial structure of the criminal, and get different witnesses to extrapolate from there. Different sketches are created from the same starting point and then you see if they match.

If they are reasonably close, you check with a photograph of a known criminal (a real world value) and see how close you are. If it matches then your starting assumptions and method of drawing were good. If not, you go back and start afresh with different assumptions and drawing techniques, over and over until something finally succeeds. It is a form of trial and error, if we are honest, but it does the trick.

Paint with All the Colours of the Quark

When you see an object's colour, what you are really detecting are vibrations in the electromagnetic field. Atoms and molecules harmonise their electrons with a certain amount of energy, which corresponds to the energy of photons being emitted or reflected.

A high-energy photon will hit your eye and your brain makes up the colour violet to account for it, while a low energy photon hits your eye and your brain makes up the colour red.

Fundamental particles do not have an actual appearance in this sense, only the photons they emit do. This is not easy to picture since most objects in our experience have some sort of colour on their surface. If you picture a tennis ball, its surface is green at a naive level but really that means the electrons in the tennis ball's surface are transferring energy into the photon field at a value your brain interprets as green.

Quarks do transfer energy into the photon field (they have electric charge) but the energies are far too high for our eyes to notice. The actual 'colour' of quarks would be the same 'colour' as a beam of X-rays or gamma rays so their appearance would be, in effect, invisible.

The same could be said for a single electron moving through space. Unless a particle actually collides with something or gets snared onto an atom and loses energy (which it emits as a photon) we would never see it coming towards us or away from us.

Electrons moving through water give off a blue glow, incidentally (a phenomenon called Cherenkov radiation) but appear slightly more purple in air (the colour of lightning) and slightly pink or green in the snow. The nucleus of an atom, and the quarks, protons and neutrons inside, are, however, utterly undetectable to the human eye.

Acknowledgements

When I was fourteen, my Science teacher, Mr Evans, gave me a textbook on quantum physics. I quickly fell in love with the topic and have wanted to write my own book about it ever since. *Fundamental* was a real labour of love and I want to thank the people who helped me realise this nerdy dream.

First and foremost I want to thank BreeAnne Kelly (who possibly loves Science even more than I do). Bree was vital in helping me structure the book, giving me critical feedback on which bits were not working and helped me get the tone right, so that the finished product was fun to read as well as fun to write.

I want to thank my partner in crime, the best writer I know, Karl Dixon. Karl gave me invaluable notes on writing style, helped me fine-tune a lot of the jokes and kept me smiling during the whole process, even at times when I did not feel like it.

Thank you to Andrew 'Hercules' Pettitt, who made crucial tweaks and polishes to the opening chapters and stopped me from getting carried away with inconsistent, waffley explanations. *Merci, mon ami!*

Thank you to Marcus Loft and Phil Pavet for making sure my physics was accurate and that my analogies did not detract from the facts.

Thank you to Becky, once again for being unendingly patient with me as I wrote this passion-project and letting me do this the way it needed to be done.

Thank you to my fearless agent Jen Christie who helped me pitch this book and for taking a chance on me . . . again.

Thank you to everyone at Robinson and Little, Brown for helping me get the book finished and for making it such an effortless process: Duncan Proudfoot for trusting me with such an ambitious topic; Amanda Keats for coordinating the mammoth task of editing; Beth Wright, my publicist, for helping me get the word out about my writing; Andy Hine and Kate Hibbert for negotiating all the international stuff; and a very special thanks to my copy-editor Howard Watson whose attention to detail is unparalleled. It has been a privilege working with you guys.

I want to thank several authors whose books were indispensable during writing. So, although I have not met these people directly, I want to thank Hagen Kleinert, Tom Lancaster, Stephen Blundell, David Tong, Anthony Zee and Leonard Susskind for helping me understand the bits I did not and feel more confident about the bits I did.

Oh, and thank you to Carly Rae Jepsen for providing the music I listened to as I wrote.

Thank you to Seishi Shimizu for being an inspiration to me as a scientist and writer.

Thank you to my father, Paul, for always believing in me. And, most importantly, thank you to everyone who went out and bought my first book which is the main reason I got to write this second one. The support from friends, family, students and strangers has been overwhelming. I hope I get to do this again for you guys!

Notes

INTRODUCTION

1. R. P. Feynman, *QED: The Strange Theory of Light and Matter* (London: Penguin, 1985).
2. S. Giles, *Theorising Modernism: Essays in Critical Theory* (London: Routledge, 1993).

CHAPTER ONE

1. A. Marmodoro, *Aristotle on Perceiving Objects* (Oxford: Oxford University Press, 2014).
2. J. Gribbin and M. Gribbin, *Science: A History in 100 Experiments* (London: William Collins, 2016).
3. E. Zalta, *Stanford Encyclopedia of Philosophy* (22 August 2017), available from: https://plato.stanford.edu/entries/descartes-physics/ (accessed 15 December 2018).
4. I. Newton, *Opticks* (1704; republished New York: Dover Publications, 1952).
5. A. Robinson, *The Last Man Who Knew Everything* (London: Oneworld Publications, 2006).
6. P. Ehrenfest, 'On the Necessity of Quanta' (1911), trans. L. Navarro and E. Perez, *Arch. Hist. Exact Sci.*, vol. 58 (2004), pp. 97–141.
7. A. Lightman, *The Discoveries* (New York: Vintage, 2006).
8. E. Cartmell and G. Fowles, *Valency and Molecular Structure* (fourth edition, London: Butterworths, 1977).

Chapter Two

1. F. Swain, *The Universe Next Door* (London: John Murray, 2017).
2. G. Lewis, 'The Conservation of Photons', *Nature*, vol. 118, no. 2981 (1926), pp. 874–5.
3. A. Howie, 'Akira Tonomura (1941–2012)', *Nature*, vol. 486, no. 7403 (2012), pp. 324.
4. N. Blaedal, *Harmony and Unity: The Life of Niels Bohr* (Lexington: Plunkett Lake Press, 2017).
5. B. Franklin, 'Experiments and Observations on Electricity', *Pennsylvania Gazette* (19 October 1752).
6. J. J. Thomson, *Recollections and Reflections* (London: G. Bell and Sons, 1936).
7. I. Asimov, *Words of Science* (London: Harrap, 1974).

Chapter Three

1. E. Wollan and L. Borst, 'Physics Section III Monthly Report for the Period Ending December 31, 1944', *Oak Ridge, Tennessee Clinton Laboratories, Metallurgical Report*, no. M-CP-2222 (1945).
2. S. Eibenberger et al., 'Matter-wave Interference with Particles Selected from a Molecular Library Masses Exceeding 10,000 amu', *Phys. Chem. Chem. Phys.*, vol. 15 (2013), pp. 14696–700.
3. D. Cassidy, 'The Sad Story of Heisenberg's Doctoral Oral Exam', *APS News*, vol. 7, no. 1 (1998).
4. J. Gribbin, *In Search of Schrödinger's Cat* (London: Transworld, 1984).
5. D. Charles, 'Heisenberg's Principles Kept Bomb From Nazis', *New Scientist*, no. 1837 (1992).
6. J. Glanz, 'Letter May Solve Nazi A-Bomb Mystery', *New York Times* (7 January 2002).
7. G. Blazeski, 'The Nazis were harassing Heisenberg, so his mother called Himmler's mom & asked her if she would please tell the SS to give her son a break', *Vintage News* (8 April 2017), available from: https://www.thevintagenews.com/2017/04/08/the-nazis-were-harassing-heisenberg-so-his-mother-called-himmlers-mom-asked-her-if-she-would-please-tell-the-ss-to-give-her-son-a-break/ (accessed 15 December 2018).

8. M. Gladwell, 'No Mercy', *New Yorker* (4 September 2006).
9. A. Trabesinger, 'The Path to Agreement', *Nature Physics*, vol. 4, no. 349 (2008).
10. D. Kevles, *The Physicists: The History of a Scientific Community in Modern America* (Cambridge, MA: Harvard University Press, 1995).
11. W. Heisenberg, *Physics and Beyond: Encounters and Conversations* (London: G. Allen & Unwin, 1971).

CHAPTER FOUR

1. W. Moore, *Schrödinger: Life and Thought* (Cambridge: Cambridge University Press, 1989).
2. Moore, *Schrödinger*.
3. E. Schrödinger, 'An Undulating Theory of the Mechanics of Atoms and Molecules', *Physical Review*, vol. 28, no. 6 (1926).
4. Moore, *Schrödinger*.
5. M. Brooks, *The Quantum Astrologer's Handbook* (Brunswick: Scribe, 2017).
6. Narcotics Anonymous, *World Service Conference of Narcotics Anonymous* (November 1981), available from: https://web.archive. org/web/20121202030403/http://www.anonymifoundation.org/ uploads/NA_Approval_Form_Scan.pdf (accessed 15 December 2018).
7. N. Camus et al., 'Experimental Evidence for Quantum Tunnelling Time', *Phys. Rev. Lett.*, vol. 119 (2017), pp. 23201.

CHAPTER FIVE

1. W. Gerlach and O. Stern, 'Der Experimentelle Nachweis der Richtungsquanteling im Magnetfeld', *Z. fur Physik*, vol. 9 (1922), pp. 349–52.

Chapter Six

1. R. Kastern, *The Transactional Interpretation of Quantum Mechanics* (Cambridge: Cambridge University Press, 2012).
2. N. D. Mermin, 'What's Wrong with this Pillow?', *Physics Today*, vol. 42, no. 4 (1989).
3. I. Born, *The Born–Einstein Letters* (New York: Walker and Company, 1971).
4. W. Heisenberg, *Physics and Beyond*, trans. A. Pomerans (New York: Harper and Row, 1971).
5. D. Lindley, *Where Does the Weirdness Go?* (New York: Vintage, 1997).
6. From a memoir of Ruth Braunizer, Erwin Schrödinger's daughter, entitled 'Memories of Dublin', collected in G. Holfter (ed.), *German Speaking Exiles in Ireland 1933–1945* (Amsterdam: Rodopi, 2006).
7. C. McDonnell, 'Schrödinger's Cat', *GITC Review*, vol. 13, no. 1 (2014).

Chapter Seven

1. J. von Neumann, *Mathematical Foundations of Quantum Mechanics*, trans. R. Bayer (Princeton: Princeton University Press, 1955).
2. E. Wigner, 'Remarks on the Mind-Body Question', in I. J. Good (ed.), *The Scientist Speculates* (London: Heinemann, 1961).

Chapter Eight

1. A. Einstein, B. Podolsky and N. Rosen, 'Can Quantum Mechanical Description of Reality be Considered Complete?', *Phys. Rev.*, vol. 47 (1935).
2. E. Schrödinger, 'Discussion of Probability Relations Between Separated Systems', *Math. Proc. of the Cam. Phil. Soc.*, vol. 31, no. 4 (1935), pp. 555–63.
3. J. E. Haynes, H. Klehr and A. Vassiliev, *Spies: The Rise and Fall of the KGB in America* (New Haven and London: Yale University Press, 2009).
4. A. Whitaker, *John Stewart Bell and Twentieth-Century Physics* (Oxford: Oxford University Press, 2016).

5. A. Aspect, P. Grainger and G. Roger, 'Experimental Realization of Einstein–Podolsky–Rosen–Bohm Gedankenexperiment: A New Violation of Bell's Inequalities', *Phys. Rev. Lett.*, vol. 49, no. 2 (1982), pp. 91–4.

CHAPTER NINE

1. R. Ji-Gang et al., 'Ground to Satellite Quantum Teleportation', *Nature*, vol. 549, no. 7670 (2017), pp. 70–3.
2. X. S. Ma et al., 'Quantum Teleportation Over 143 Kilometers Using Active Feed-forward', *Nature*, vol. 489, no. 7415 (2012), pp. 269–73.
3. C. Bennett et al., 'Teleporting an Unknown Quantum State via Dual Classical and Einstein–Podolsky–Rosen Channels', *Phys. Rev. Lett.*, vol. 70, no. 13 (1993), pp. 1895–9.
4. P. Ball, 'Quantum Teleportation is Even Weirder Than You Think', *Nature Column: Muse* (20 July 2017).
5. Y. H. Kim et al., 'A Delayed Choice Quantum Eraser', *Phys. Rev. Lett.*, vol. 84 (2000), pp. 1–5.
6. C. Marletto et al., 'Entanglement Between Living Bacteria and Quantized Light Witnessed by Rabi Splitting', *Journal of Phys. Comm.*, vol. 2, no. 40 (2018).

CHAPTER TEN

1. W. Keepin, 'Lifework of David Bohm' (11 March 2008), available from: http://www.vision.net.au/~apaterson/science/david_bohm.htm (accessed 15 December 2018).
2. Y. Couder et al., 'Walking Droplets: a Form of Wave-particle Duality at Macroscopic Level?', *Europhys. News*, vol. 41, no. 1 (2010), pp. 14–18.
3. J. Bush et al., 'Walking Droplets Interacting with Single and Double Slits', *J. Fluid Mech.*, vol. 835 (2018), pp. 1136–56; T. Bohr et al., 'Double Slit Experiment with Single Wave-driven Particles and Its Relation to Quantum Mechanics', *Phys. Rev. E.*, vol. 92 (2015).

4. J. Cramer, 'The Transactional Interpretation of Quantum Mechanics and Quantum Nonlocality' (2015), available from: https://arxiv.org/pdf/1503.00039.pdf (accessed 15 December 2018).

5. D. Deutsch, *The Beginning of Infinity* (London: Penguin, 2012).

6. F. Tipler, *The Physics of Immortality* (New York: Bantam Doubleday Dell Publishing Group, 2000).

7. P. Byrne, *The Many Worlds of Hugh Everett III: Multiple Universes, Mutual Assured Destruction and the Meltdown of a Nuclear Family* (Oxford: Oxford University Press, 2010).

8. P. Ball, 'Experts Still Split about What Quantum Theory Means', *Nature News* (11 January 2013). The original poll can be found at: https://arxiv.org/pdf/1301.1069.pdf (accessed 15 December 2018).

9. I. Asimov, 'Science and the Bible', interview with Prof. Asimov conducted by P. Kurtz in *Free Enquiry*, Spring (1982).

CHAPTER ELEVEN

1. I. Asimov, *New Guide to Science* (Harmondsworth: Penguin Press Science, 1993).

2. G. Farmelo, *The Strangest Man: The Hidden Life of Paul Dirac, Quantum Genius* (London: Faber and Faber, 2009).

3. P. Dirac, *Lectures on Quantum Mechanics* (New York: Dover Publications, 2001).

CHAPTER TWELVE

1. C. Sykes, *No Ordinary Genius* (London: W. W. Norton & Company, 1994).

2. J. Gleick, *Genius* (London: Little, Brown, 1992).

3. Letter from Robert Oppenheimer addressed to Robert Birge, dated 4 November 1943.

4. R. Leighton, *Surely You're Joking, Mr Feynman* (Princeton: Princeton University Press, 1985).

5. A. Zee, *Quantum Field Theory in a Nutshell* (Princeton: Princeton University Press, 2010).

6. Sykes, *No Ordinary Genius*.

7. M. Nio et al., 'Complete Tenth-order QED Contribution to the Muon g-2' (2012), available from: https://arxiv.org/abs/1205.5370 (accessed 15 December 2018).

8. T. Lancaster and S. Blundell, *Quantum Field Theory for the Gifted Amateur* (Oxford: Oxford University Press, 2015).

9. R. P. Feynman, 'The Theory of Positrons', *Phys. Rev.*, vol. 76 (1949).

10. C. D. Anderson, 'The Positive Electron', *Phys. Rev.*, vol. 43 (1933).

11. D. Dooling, 'Reaching for the Stars', *Science at NASA* (12 April 1999), available from: https://science.nasa.gov/science-news/science-at-nasa/1999/prop12apr99_1 (accessed 15 December 2018).

12. W. Bertsche et al., 'Confinement of Antihydrogen for 1,000 seconds', *Nature Phys.*, vol. 7, no. 7 (2011), pp. 558–64.

CHAPTER THIRTEEN

1. Author Unknown, 'Who Ordered That?', *Nature Editorial*, vol. 531 (2016), pp. 139–40.

2. M. L. Perl et al., 'Evidence for Anomaloys Lepton Production in e+ e– Annihilation', *Phys. Rev. Lett.*, vol. 35, no. 22 (1975).

3. R. P. Feynman, *QED: The Strange Theory of Light and Matter* (London: Penguin, 1985).

4. M. Kaku, 'Beauty Is Truth', *Forbes Magazine* (7 October 2008).

5. L. Lederman, 'Neutrino Physics', Lecture given on 9 January 1963, *Brookhaven Lecture Series on Unity of Science*, BNL 787, no. 23.

6. Interview with Gell-Mann, available from: https://www.youtube.com/watch?v=po-SQ33Kn6U (accessed 15 December 2018).

7. F. Wilczek, 'Time's (Almost) Reversible Arrow', *Quanta Magazine* (7 January 2016).

8. M. E. Peskin and D. V. Schoeder, *An Introduction to Quantum Field Theory* (Boston: Addison-Wesley, 1995).

CHAPTER FOURTEEN

1. K. Jepsen, 'Famous Higgs Analogy, Illustrated', *Symmetry Magazine* (9 June 2013).
2. J. W. Brewer and M. K. Smith, *Emmy Noether: A Tribute to Her Life and Work* (New York: Marcel Dekker Inc., 1981).
3. A. Einstein, 'Obituary of Amalie "Emmy" Noether', *New York Times* (5 May 1935).
4. Author Unknown, 'Why Is the Higgs Discovery so Significant?', *Science and Technology Facilities Council* (22 September 2017), available from: https://stfc.ukri.org/research/particle-physics-and-particle-astrophysics/peter-higgs-a-truly-british-scientist/why-is-the-higgs-discovery-so-significant (accessed 15 December 2018).
5. P. Rogers, 'The Heart of the Matter', *Independent* (1 September 2004).
6. L. Lederman, *The God Particle* (New York: Dell, 1993).

CHAPTER FIFTEEN

1. W. Stukeley, *Memoirs of Sir Isaac Newton's Life* (1752; republished London: The Royal Society, 2010).
2. A. D. Aczel, *God's Equation* (New York: Delta, 2000).

Index